T0341017

Internet of Things
and M2M Communications

RIVER PUBLISHERS SERIES IN INFORMATION SCIENCE AND TECHNOLOGY

Consulting Series Editor

KWANG-CHENG CHEN
National Taiwan University
Taiwan

Information science and technology enables 21st century into an Internet and multimedia era. Multimedia means the theory and application of filtering, coding, estimating, analyzing, detecting and recognizing, synthesizing, classifying, recording, and reproducing signals by digital and/or analog devices or techniques, while the scope of "signal" includes audio, video, speech, image, musical, multimedia, data/content, geophysical, sonar/radar, bio/medical, sensation, etc. Networking suggests transportation of such multimedia contents among nodes in communication and/or computer networks, to facilitate the ultimate Internet. Theory, technologies, protocols and standards, applications/services, practice and implementation of wired/wireless networking are all within the scope of this series. We further extend the scope for 21st century life through the knowledge in robotics, machine learning, cognitive science, pattern recognition, quantum/biological/molecular computation and information processing, and applications to health and society advance.

- Communication/Computer Networking Technologies and Applications
- Queuing Theory, Optimization, Operation Research, Statistical Theory and Applications
- Multimedia/Speech/Video Processing, Theory and Applications of Signal Processing
- Computation and Information Processing, Machine Intelligence, Cognitive Science, and Decision

For a list of other books in this series, please visit www.riverpublishers.com

Internet of Things and M2M Communications

Editors

Fabrice Theoleyre
CNRS, University of Strasbourg, France

and

Ai-Chun Pang
National Taiwan University, Taiwan

Routledge
Taylor & Francis Group

LONDON AND NEW YORK

Published 2013 by River Publishers
River Publishers
Alsbjergvej 10, 9260 Gistrup, Denmark
www.riverpublishers.com

Distributed exclusively by Routledge
4 Park Square, Milton Park, Abingdon, Oxon OX14 4RN
605 Third Avenue, New York, NY 10158

First published in paperback 2024

Internet of Things and M2m Communications / by Fabrice Theoleyre, Ai-Chun Pang.

Routledge is an imprint of the Taylor & Francis Group, an informa business

Publisher's Note
The publisher has gone to great lengths to ensure the quality of this reprint but points out that some imperfections in the original copies may be apparent.

While every effort is made to provide dependable information, the publisher, authors, and editors cannot be held responsible for any errors or omissions.

ISBN: 978-87-92982-48-3 (hbk)
ISBN: 978-87-7004-511-7 (pbk)

Table of Contents

Part III. Security & Tests

Preface

The Internet of Things (IoT) has emerged because of the wide generalization of wireless communications and more and more energy efficient radio chipsets. In recent years, it has attracted much attention from the academic and industrial researchers. The IoT aims at introducing smartness in our common life: smart objects are interconnected to the Internet and may communicate directly with each other. Such communication paradigm triggers innovative applications.

For this book on IoT, we selected the best papers from the 15th International Symposium on Wireless Personal Multimedia Communications that took place in September 2012 in Taiwan. We also invited some open submissions to cover unfilled topics in the M2M area.

The IoT integrates smart objects using small batteries. Thus, IoT must integrate the energy constraint in the whole networking stack. The first part of the book focuses on the adaptations of the networking layers to this specific constraint.

In the first part, Chapter 1 explores how the networking stack must be modified to be *data-aware*: concatenating the data packets may help to improve energy savings. The authors introduce an analytical model based on a Markov chain, and then discuss on the effect of data aggregation in M2M communications. Chapter 2 investigates the impact of opportunistic routing in energy-harvesting networks. Opportunistic routing helps to select the most accurate next hop, based on the real instantaneous radio conditions. Furthermore, the authors also propose to adapt the duty-cycle dynamically to cope efficiently with energy-harvesting nodes. Chapter 3 proposes an offline-tool to estimate the energy consumption of the PHY layer. In this way, the network designer is able to validate *a priori* the deployment of its solution.

The second part of the book concentrates attention on the energy constraints of M2M communications. Chapter 4 investigates the problem of Delay Constrained Scheduling from a queuing theory point of view. The authors detail how an optimal scheduling can be achieved while guaranteeing a minimum packet delivery ratio under a delay constraint. The IoT may enable the tracking of mobile objects. Chapter 5 studies how distributed clustering and a proper scheduling allows the network to track any mobile objects. The authors apply Bayesian filters to accurately estimate the location of the targets. Chapter 6 investigates the synchronization problem in multichannel wireless networks. Indeed, synchronization is often required for, e.g., medium access or sleeping schedule. The authors present a bio-inspired synchronization solution.

Finally, the last part of the book focuses on the security concerns and validation of the solutions for the IoT. Chapter 7 explores how we may secure the control access in M2M communications. Mechanisms permit to ensure both confidentiality and authentication. Since the IoT will aggregate several service providers, the authors also explain how a delegation between different domains is achievable. Chapter 8 investigates how to combat eavesdroppers. The authors use normally inactive relays in the selective decode and forward cooperative transmission. This cooperative jamming helps to improve transmission secrecy. To evaluate properly the behavior of M2M applications, the authors propose, in Chapter 9, to execute the applications in virtual machines. The authors present a whole architecture to efficiently evaluate any M2M application.

Acknowledgements

We would like to thank Kwang-Cheng Chen, the general chair of the International Symposium on Wireless Personal Multimedia Communications 2012. We much appreciate his time, advice and guidance in assisting us with the publication of this book on M2M communications. We also thank the organization committee members for their help. Importantly, we wish to thank the team of international reviewers and the Technical Program Committee who freely gave their time to provide a constructive feedback that is important to judge and improve the quality of papers. Finally, we would like to thank River Publishers team for their editorial guidance.

Fabrice Théoleyre & Ai-Chun Pang

Part I

Energy Constrained IoT

1

Effect of Data Aggregation in M2M Networks

Shin-Yeh Tsai[1], Sok-Ian Sou[1] and Meng-Hsun Tsai[2]

[1]*Institute of Computer and Communication Engineering, National Cheng Kung University, Tainan, Taiwan, R.O.C.*
e-mail: shinyehtsai@gmail.com, sisou@mail.ncku.edu.tw
[2]*Department of Computer Science and Information Engineering, National Cheng Kung University, Tainan, Taiwan, R.O.C.*
e-mail: tsaimh@csie.ncku.edu.tw

Abstract

Machine-to-Machine (M2M) communication technologies provide capabilities for devices to communicate with each other through wired and wireless
system. Applying data aggregation is an efficient way to reduce energy
consumption of M2M network. In this chapter, we first introduce data aggregation, and devise an analytical model to compute the transmission delay and
energy efficiency in data aggregation. Then we develop an extensive simulation to validate our proposed analytical model. Numerical results show that
an increment of buffering time results in a significant decrement in energy
consumption. After the simulation, we point out that routing algorithms, timing strategies and security issue are the open challenges of data aggregation
in the near future. This chapter provides guidelines to configure the buffering
time for data aggregation in M2M networks.

Keywords: buffering time, data aggregation, energy efficiency, M2M.

Fabrice Theoleyre and Ai-Chun Pang (Eds.), Internet of Things and
M2M Communications, 3–22.

1.1 Introduction

Machine-to-Machine (M2M) communication technologies provide capabilities for devices to communicate with each other through wired and wireless systems. Major M2M applications include alerting and habitat monitoring. Consider the scenario illustrated in Figure 1.1, where data is collected at M2M devices such as monitoring devices (see Figure 1.1(a)). Then, M2M devices deliver data to an M2M application server (see Figure 1.1(c)) via an M2M gateway (see Figure 1.1(b)).

As the scale of M2M networks becomes extremely large, a significant amount of transmission overhead will be observed at M2M devices. Since the sensing data can be retrieved from the same kind of M2M devices, reducing transmission overhead by data aggregation is an efficient way to improve the energy efficiency of M2M network [6].

To utilize data aggregation, M2M devices need to exercise timing strategies to determine the length of an aggregation period. The study in [11] points out that the timing strategy has significant performance impacts on data accuracy and freshness. With improper timing strategies, poor performance in terms of delivery delay and energy efficiency would be observed. The key factor of the timing strategy of data aggregation is how to determine buffering timer precisely. To address this issue, we propose an analytical model to investigate the effect of buffering time in aggregation mechanisms.

The rest of this chapter is organized as follows. Section 1.2 presents the related works on data aggregation in M2M networks. Section 1.3 describes the system architecture of packet buffering in aggregation mechanisms. Section 1.4 proposes an analytical model to derive mechanism output metrics in terms of the delivery delay, aggregated volume and energy consumption. Section 1.5 develops simulations to study the numerical examples. Section 1.6 discusses open challenges in data aggregation. Finally, Section 1.7 concludes this chapter.

1.2 Related Works

A wealth of work studies on data aggregation in recent years. There are three significant issues in data aggregation. The first issue is about choosing aggregation points; the second issue is the routing algorithm in data aggregation; and the third issue is about the timing strategy. However, in [2], Fasolo points out that most of the current studies mainly focus on the first two issues, while

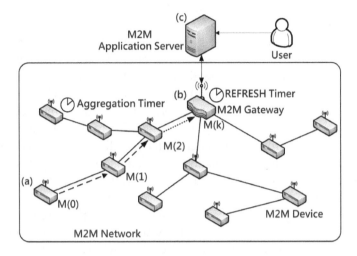

Figure 1.1 An architecture for M2M network.

only a few studies provide deeper analysis on the timing strategies in data aggregation.

In [6], Kalpakis proposes a basic network model for discussing the effects of data aggregation and data gathering. In Kalpakis's network model, each device periodically monitors information and delivers the information hop by hop to the core network (e.g., base stations, sinks and M2M servers; see Figure 1.1(c)) for later analysis. Kalpakis points out that, in a deployment of energy-constrained devices, data aggregation and data gathering can prolong the lifetime of total network.

In such energy-constrained environment, there are two major types of routing algorithms in data aggregation. The first type is cluster-based approaches. The study in [5] proposes a clustering protocol which is named Low-Energy Adaptive Clustering Hierarchy (LEACH) to improve the network lifetime. In order to let the energy consumption on each device evenly distribute, LEACH utilizes randomization to pick cluster heads. Each device keeps one-hop distance with its cluster head. After gathering information from cluster members, cluster heads aggregate and deliver the gathered information to core network. LEACH is a self-organizing protocol which does not need to construct routing topology through global information or control message from core network. Furthermore, LEACH provides flexibility of aggregation strategies which are implemented on cluster heads. However, in a high mobility environment, the costs to maintain a stable cluster will be extremely high.

Cougar [15] is another cluster-based routing algorithm. Unlike the randomization in LEACH, Cougar selects cluster heads based on signal strength. Moreover, there is no limitation on the distance between cluster members and the cluster head. The distance can be more than one-hop. Cluster members utilize Ad Hoc On Demand Distance Vector to hold communication with their cluster heads. Each cluster head owns a list of its members. After receiving the packets from the members in the list, the cluster head delivers the gathered and aggregated information. The weakness of Cougar is the same as the weakness of LEACH. Cougar is also hard to manage the environment which is highly mobile.

The second type is tree-based approach. A spanning tree is constructed from the core network to the end devices. All the gathered information is routed through the tree structure. The Tiny AGgregation (TAG) [8] is a well-known tree-based routing algorithm. The implementation of TAG is composed by two parts: (1) *Distribution Phase* and (2) *Collection Phase*. During distribution phase, control messages are exchanged between core network and end devices to configure the setting of aggregation, and to construct the spanning tree. During collection phase, all the intermediate devices perform data aggregation. After all the periodic data from children is gathered, parents deliver the aggregated information through the routing tree which is established during distribution phase.

The study in [9] proposes an energy-aware routing algorithm, which routes packets based on the remaining energy in each device. Mottola shows that data aggregation improves up to 80% of lifetime for whole network through real world experiments. However, the side effect of aggregation is significant, where the transmission delays with aggregation increase about four times. Besides, the authors [9] compare only the performance of routing algorithm with aggregation to that without aggregation. The timing strategies are not investigated in the real world experiments of [9].

The study in [11] points out that when to clock out the buffering timer has significant performance impacts on data accuracy and freshness. Solis studies two timing strategies in data aggregation: (1) *Periodic Simple* and (2) *Periodic Per-hop*. In *periodic simple* mechanism, packets are buffered until the expiration of a buffering timer. Upon expiration of the buffering timer, an M2M device retrieves the data which is gathered in the buffering period and aggregates them into a new packet. In *periodic per-hop* mechanism, packets are buffered until the number of buffered packets reaches a threshold. *Periodic per-hop* mechanism further combines a buffering timer on a packet counter to avoid overlong buffering time. Upon excess of maximum buffered packets or expiration of the buffering timer, the device aggregates the data into a new packet. In this chapter, we mainly focus on the *periodic simple* mechanism.

When we evaluate the efficiency of data aggregation, data accuracy, latency, energy efficiency and network lifetime are four key factors to determine it [10]. However, the evaluation metrics of energy efficiency in M2M networks and wireless sensor networks are significant different. Lifetime is the major metric which we use to evaluate the energy efficiency in wireless sensor networks. In wireless sensor networks, study in [3] further proposes a comprehensive concept of calculation on lifetime (First Node Dies, Half of the Nodes Alive and last Node Dies). However, in M2M networks, with directly power supply, there is no lifetime critical. In M2M networks, the total energy consumption is the major evaluation metric of energy efficiency. In this chapter, we focus on the performance metrics concerning the respective latency to the application server, the aggregated volume and energy consumption against the buffering time configured in the M2M devices.

1.3 System Model

This section describes the *periodic simple* mechanism in M2M network. In Figure 1.1, each M2M device (Figure 1.1(a)) collects sensing data in its located area. The collected data is then sent to an M2M application server (Figure 1.1(c)) via an M2M gateway (Figure 1.1(b)). As shown in Figure 1.1, the source nodes and the M2M gateway are denoted by $M(j)$, $0 \le j \le k-1$, and $M(k)$, respectively. The data e_j sensed by $M(j)$ is delivered to the application server through the routing path $M(j) \to M(j+1) \to \cdots \to M(k)$. *Periodic simple* mechanism is implemented at each M2M device $M(i)$, $0 \le i \le k - 1$. It works as follows:

Step 1. When $M(i)$ receives a packet containing the sensing data e_j, $0 \le j \le i - 1$, from $M(i - 1)$ or generates a packet containing the data e_i which is sensed by $M(i)$ itself, $M(i)$ puts the packet in its aggregation buffer.

Step 2. If the aggregation buffer is empty, $M(i)$ starts an aggregation timer of period T.

Step 3. When the aggregation timer expires, $M(i)$ retrieves the buffered packets and aggregates the collected sensing data into a new packet.

Note that the aggregated packet contains data e_j and e_i. Then, $M(i)$ sends the new packet to $M(i + 1)$.

After the sensing data e_i, $0 \le i \le k - 1$, arrives at the M2M gateway $M(k)$, $M(k)$ updates the sensing data stored in the application server. To reduce signal overhead in the core network, the M2M gateway may update the M2M server after a time period denoted by τ. This special technique is called delayed REFRESH operation, which has been widely used in 3GPP core network [1, 12].

1.4 Analytical Model

This section proposes an analytical model to compute the following perform-ance metrics:

- Expected delivery delay $E[t_s]$: the average time period between when data is sensed at the source and when data is received at the M2M application server.
- Expected aggregated volume $E[N]$: the expected number of sensing data aggregated in a packet.
- Expected energy consumption $E[E_T]$: the expected energy consumption on a device in a unit time.
- Expected number $E[N_R]$ of REFRESH operations in an aggregation period.

Let t_r and t_a be the inter REFRESH interval and the inter packet arrival interval with mean $1/\lambda_R$ and $1/\lambda_A$, respectively. Suppose that the aggregation timer T has mean $1/\lambda_T$, and τ_T denotes the residual time of T. We assume that t_r, t_a and T are exponentially distributed.

1.4.1 Delivery Delay

Figure 1.2 illustrates the timing diagram between two REFRESH operations which occur at time t_0 and time $t_0 + t_r$, respectively. These two consecutive REFRESH operations are triggered by an M2M gateway. The delivery delay between $M(j)$, $0 \le j \le k - 1$, and the M2M server is denoted by t_s. The data e_j sensed by $M(j)$ arrives at devices $M(i)$ at time $t_{M,i}$, $j + 1 \le i \le k$. The transmission delay between $M(i - 1)$ and $M(i)$ is denoted by δ_i. Let t_g

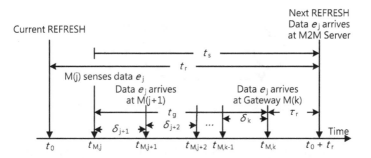

Figure 1.2 Timing diagram to illustrate the delivery status.

denote the transmission delay between $M(j)$ and $M(k)$, then

$$t_g = t_{M,k} - t_{M,j} = \sum_{i=j+1}^{k} \delta_i \qquad (1.1)$$

When a packet arrives at $M(i)$, there are two cases: (1) the aggregation buffer is empty and (2) the aggregation buffer is not empty. In Case (1), $M(i)$ starts a buffering timer of period T and sends the packet when the timer expires. In Case (2), buffering timer is already started in $M(i)$. Therefore, $M(i)$ sends the packet when the started timer expires. The delay time δ_i is expressed as

$$\delta_i = \begin{cases} T, & \text{Case (1)} \\ \tau_T, & \text{Case (2)} \end{cases}$$

Since the buffering time T is exponentially distributed, based on the memoryless property, the residual time τ_T of the timer has the same distribution as T. Therefore, the delay time δ_i has the same distribution as T with mean $1/\lambda_T$. Assume that delay time δ_i experienced in the ith hop transmission has density function $f_\delta(\cdot)$, then

$$f_\delta(\delta_i) = \lambda_T e^{-\lambda_T \delta_i} \qquad (1.2)$$

From Eqs. (1.1) and (1.2), t_g has Erlang distribution with mean $(k - j)/\lambda_T$, and variance $(k - j)/\lambda_T^2$. The density function $f_g(\cdot)$ of t_g is expressed as

$$f_g(t_g) = \frac{\lambda_T^{k-j}(t_g)^{k-j-1}e^{-\lambda_T t_g}}{(k - j - 1)!} \qquad (1.3)$$

Let $\tau_r = t_0 + t_r - t_{M,i}$ be the residual period of REFRESH operation (i.e., the interval between packet generation and the next REFRESH operation). From

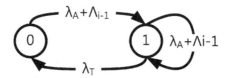

Figure 1.3 The state transition rate diagram for the device buffer state model.

the memoryless property, τ_r is also exponentially distributed. Therefore, the density function $f_R(\cdot)$ is expressed as

$$f_R(\tau_r) = \lambda_R e^{-\lambda_R \tau_r} \tag{1.4}$$

The expected delivery delay $E[t_s]$ is composed by the transmission delay t_g and τ_r after the packet arrives at the M2M gateway $M(k)$. Since τ_r has the same distribution with t_r, we have $E[\tau_r] = E[t_r]$. Therefore, from Eqs. (1.3) and (1.4), we have

$$
\begin{aligned}
E[t_s] &= E[t_g] + E[\tau_r] \\
&= \frac{k-j}{\lambda_T} + \frac{1}{\lambda_R}
\end{aligned}
$$

1.4.2 Aggregated Volume

The packets processed by $M(i)$ are classified into two types: Type 1 packet is generated by $M(i)$ itself; Type 2 packet is generated by $M(j)$, $j \neq i$. The packet arrival rate of Type 1 packet is λ_A, and the packet arrival rate of Type 2 packet is equal to the packet departure rate of $M(i-1)$. When $M(i)$ has a higher packet arrival rate, the packet departure rate of $M(i)$ increases as well. Let Λ_i be the packet departure rate of device $M(i)$, $0 \leq i \leq k-1$. Let $t_{a,i}$ denote the inter packet arrival interval at device $M(i)$. We define a continuous-time Markov process $X(t)$ to model the buffer state of a device, where

$$X(t) = \begin{cases} 1, & \text{buffer is not empty} \\ 0, & \text{buffer is empty} \end{cases}$$

The event of packet delivery is occurred when the buffer state changes from 1 to 0. Figure 1.3 shows the state transition rate diagram for the Markov process $X(t)$. Since the inter packet arrival interval $t_{a,i}$ and expiration of timer T are exponentially distributed with mean $1/(\lambda_A + \Lambda_{i-1})$ and $1/\lambda_T$, respectively, we have

$$P[X(t) = 0](\lambda_A + \Lambda_{i-1}) = P[X(t) = 1]\lambda_T \tag{1.5}$$

By solving Eq. (1.5) with $P[X(t) = 0] + P[X(t) = 1] = 1$, Λ_i can be expressed by $P[X(t) = 1]$ and the expiration of buffering time as

$$\begin{aligned}
\Lambda_i &= P[X(t) = 1]\lambda_T \\
&= \frac{(\lambda_A + \Lambda_{i-1})\lambda_T}{\lambda_A + \Lambda_{i-1} + \lambda_T}
\end{aligned} \tag{1.6}$$

where $\Lambda_{-1} = 0$ and $\Lambda_k = 0$.

Let P_{α_i} be the probability that, the packet arriving at $M(i)$ is of Type I while the aggregation buffer of $M(i)$ is empty. We have

$$P_{\alpha_i} = \frac{\lambda_A}{\lambda_A + \Lambda_{i-1}} \tag{1.7}$$

The aggregated volume of Type I packet is equal to 1, and the aggregated volume of Type II packet at $M(i)$ is equal to the aggregated volume of the departure packet of $M(i-1)$ (i.e., $E[N_{i-1}]$). Since the packets arrive at $M(i)$ in an buffering period can be expressed as $(\lambda_A + \Lambda_{i-1}E[N_{i-1}])/\lambda_T$, from Eqs. (1.6) and (1.7), $E[N_i]$ is expressed as

$$\begin{aligned}
E[N_i] &= P_{\alpha_i} + (1 - P_{\alpha_i})E[N_{i-1}] + \frac{\lambda_A + \Lambda_{i-1}E[N_{i-1}]}{\lambda_T} \\
&= \frac{\lambda_A + \Lambda_{i-1}E[N_{i-1}]}{\lambda_A + \Lambda_{i-1}} + \frac{\lambda_A + \Lambda_{i-1}E[N_{i-1}]}{\lambda_T}
\end{aligned}$$

when $\lambda_T = 0$, $E[N_i] = 1$.

1.4.3 Energy Consumption

The energy consumption on a device depends on the arrival rate and departure rate of the device. When a device $M(i)$ has a higher packet arrival rate, the departure rate and energy consumption of $M(i)$ increases as well. Let E_D and E_R denote the energy consumption on a single delivery and reception, respectively. The details of E_D and E_R are illustrated in Section 1.5. Let $E_T^{(i)}$ be the total amount of energy consumed in device $M(i)$, $0 \le i \le k-1$. Then, $E_T^{(i)}$ can be expressed as

$$E_T^{(i)} = \Lambda_{i-1}E_R + \Lambda_i E_D$$

where Λ_i $(0 \le i \le k - 1)$ can be obtained through recursive computation of Eq. (1.6).

1.4.4 The Expected Number of REFRESH Operations in an Aggregation Period

In order to investigate the relationship between REFRESH and buffering time, the assumption of exponential distribution on buffering time is relaxed in this subsection. We consider that the transmission delay t_g has general density function $f_g(t_g)$ and Laplace transform $f_g^*(g)$. Then the probability mass function of the number $N(t_g)$ of REFRESH operation can be computed by

$$\Pr[N(t_g) = m] = \left[\frac{(\lambda_R t_g)^m}{m!}\right] e^{-\lambda_R t_g} \tag{1.9}$$

Based on the above assumptions, from Eq. (1.9), the number N_R of REFRESH operation in time period $(t_{M,j}, t_{M,k})$ is derived as

$$\begin{aligned}
\Pr[N_R = m] &= \int_{t_g=0}^{\infty} \Pr[N(t_g) = m] f_g(t_g) dt_g \\
&= \int_{t_g=0}^{\infty} \left[\frac{(\lambda_R t_g)^m}{m!}\right] e^{-\lambda_R t_g} f_g(t_g) dt_g \\
&= \left[\frac{(-\lambda_R)^m}{m!}\right] \left[\frac{d^m f_g^*(g)}{dg^m}\right]\bigg|_{g=\lambda_R}
\end{aligned} \tag{1.10}$$

Note that our derivation can be applied to any transmission delay distribution whose Laplace Transform has close form. For demonstration purpose, we consider that t_g has Gamma distribution with mean m_s and variance v_s. The Gamma distribution is considered because the distribution of any positive random variable can be approximated by a mixture of Gamma distributions (see Lemma 3.9 in [7]). The Laplace transforms of transmission delay $f_g^*(g)$ is

$$f_g^*(g) = \left[1 + \left(\frac{v_g}{m_g} g\right)\right]^{-\frac{m_g^2}{v_g}} \tag{1.11}$$

Substituting Eq. (1.11) into Eq. (1.10) to yield

$$\Pr[N_R = m] = \frac{1}{m!} \left(\frac{-\lambda_R v_g}{m_g}\right)^m \prod_{i=0}^{m-1} \left(\frac{m_g^2}{v_g} - i\right) \left(1 + \frac{v_g}{m_g} \lambda_R\right)^{-\frac{m_g^2}{v_g} - m} \tag{1.12}$$

From Eq. (1.12), we can compute the expected number of REFRESH in aggregation period as

$$E[N_R] = \sum_{m=0}^{\infty} m \Pr[N_R = m]$$

1.5 Numerical Example

We develop a discrete event-based simulation similar to the one in [13], and the details are omitted. The above analytical model is validated against the simulation. Based on the validated simulation, this section investigates the performance of the aggregation mechanism. To observe energy efficiency in data aggregation, we measure energy consumption based on the energy model given in [4]. Let E_D and E_R denote the energy consumption on delivery and reception, respectively, then

$$\begin{cases} E_D &= l E_{elec} + l \epsilon_{amp} d^2 \\ E_R &= l E_{elec} \end{cases}$$

where E_{elec} is the cost of running transmitter or receiver circuitry and ϵ_{amp} is the cost of running transmit amplifier. We consider the free space model (d^2 power loss) in channel transmission. The packet length is denoted by l. Assume that the energy consumed in a delivery is amortized among all aggregated packets. Let N be the amount of packets to amortize energy consumption in a delivery. At each hop, let $E_W^{(n)}$ be the total amount of energy consumed from a source to the gateway of the packet which is generated by the nth-hop device, $0 \le n \le k$. Then, $E_W^{(n)} = \sum_{j=0}^{n}(E_D + E_R)/N$.

In simulation experiments, we assume that there are nine devices (i.e., $k = 8$), in the routing path and the packet arrival rate λ_A is $50\lambda_R$. The distance between each device is 100 meters. Based on the device information given by [14], we set the packet length of each packet to 29 bytes.

1.5.1 Effects of Mean Buffering Time

Figure 1.4(a) depicts the effect of $E[T]$ on the expected delivery delay $E[t_s]$. We note that no aggregation is exercised when T is set to 0. As $E[T]$ increases, $E[t_s]$ increases. At $M(1)$, as $E[T]$ increases from 0 to $0.02/\lambda_R$, delivery delay increases as well (i.e., from $1/\lambda_R$ to $1.13/\lambda_R$). This phenomenon follows our intuition that the longer a packet is buffered, the later the packet arrives at the application server. Figure 1.4(b) illustrates the impact of $E[T]$

on $E[N]$. The figure indicates that $E[N]$ increases as $E[T]$ increases. When a device increases buffering time, more packets are likely to arrive at the device in a single aggregation period. At $M(1)$, as $E[T]$ increases from 0 to $0.02/\lambda_R$, aggregated volume increases as well (i.e., from 1 to 3.27). As shown in Figure 1.4, the discrepancies between the analytical results and the simulation results are within 1%.

In Figure 1.5(a), as $E[T]$ increases, $E_W^{(n)}$ decreases. The figure shows that $E_W^{(n)}$ significantly decreases with small $E[T]$ values. In other words, compared to exercise without aggregation (i.e., $E[T] = 0$), setting a small buffering time (e.g., $E[T] = 0.02/\lambda_R$) is sufficient to significantly reduce energy consumption on a packet. Figure 1.5(b) depicts the impacts of $E[T]$ on $E_T^{(i)}$. The figure shows that energy consumption on a device $E_T^{(i)}$ significantly decreases when devices implement aggregation mechanism. When $E[T] \geq 0.02/\lambda_R$, the value of $E_T^{(i)}$, $i = \{1, 3, 5, 7\}$ are almost the same, which is significant from that in Figure 1.5(b). The reason of this convergence is that when devices implement the same buffering time, the packet departure rate of each device is almost the same. The discrepancy between the analytical results and simulation results are within 1%. In Figures 1.4(a) and 1.5, compared to exercise without aggregation, when $E[T]$ equals 0.02, $E_W^{(1)}$ and $E_T^{(1)}$ decrease about 88% and $E[t_s]$ of $M(1)$ only increases about 13%. The reason of the difference between these two gaps is that if each packet can wait for a short time, other packets can arrive at this device in this buffering interval. Even if no packet arrives before the expiration of the buffering timer, the device only buffers the packet for a short time without incurring a long delay. If packets arrive at the device before the expiration, the energy consumption of each packet is reduced. Indeed, it is worthwhile to wait for a short time for another packet to share the energy consumption in a packet delivery.

1.5.2 Effects of Mean and Variance of Inter Packet Arrival Interval

We analyze the output metrics against the variance of inter packet arrival interval. For illustration purpose, we only observe the delivery delay and energy consumption of $M(4)$ (i.e., the device in the middle of the path). We assume that t_a has the Gamma distribution with mean $1/\lambda_A$ and variance $V[t_a]$. In this subsection, we assume that the packet buffering time $E[T]$ is $1/50\lambda_R$. As $V[t_a]$ increases, more long inter packet arrival intervals are observed.

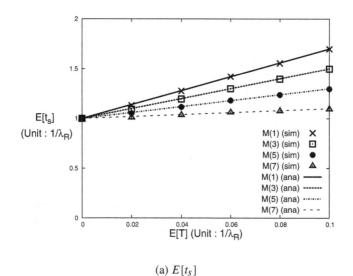

(a) $E[t_s]$

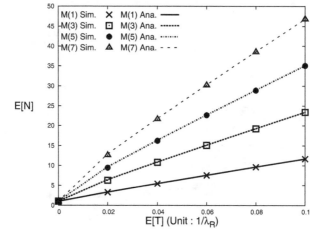

(b) $E[N]$

Figure 1.4 Effects of $E[T]$ and n on $E[t_s]$ and $E[N]$ ($\lambda_A = 50\lambda_R$).

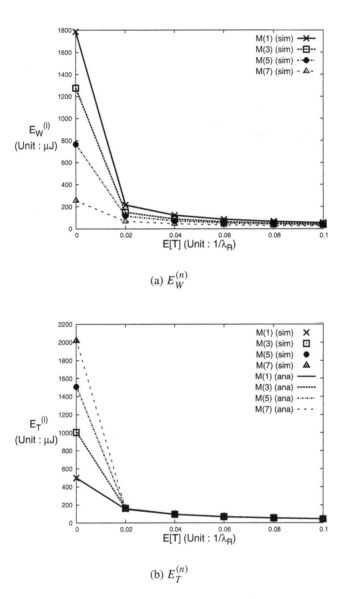

(a) $E_W^{(n)}$

(b) $E_T^{(n)}$

Figure 1.5 Effects of $E[T]$ and n on $E_W^{(n)}$ and $E_T^{(n)}$ ($\lambda_A = 50\lambda_R$).

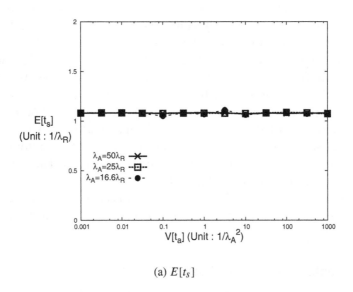

(a) $E[t_s]$

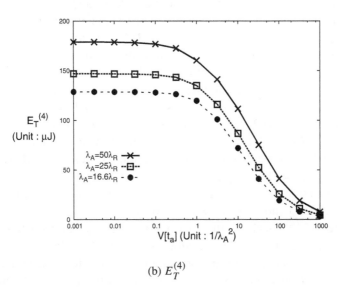

(b) $E_T^{(4)}$

Figure 1.6 Effects of λ_A and $V[t_a]$ $(E[T] = 1/50\lambda_R)$.

Figure 1.6(a) shows that $E[t_s]$ is not sensitive to the mean and variance of inter packet arrival interval. No matter what $V[t_a]$ and λ_A increase or decrease, $E[t_s]$ remains invariant. The reason of this phenomenon is that *periodic simple* mechanism only considers with buffering time. Therefore, the mean and variance of packet arrival time does not affect $E[t_s]$.

Figure 1.6(b) illustrates the effects of $V[t_a]$ and λ_A on $E_T^{(4)}$. Long inter packet arrival intervals result in smaller energy consumption. When $\lambda_A = 50\lambda_R$, as $V[t_a]$ increases from $0.01/\lambda_A^2$ to $1000/\lambda_A^2$, there is a significant decrement on energy consumption (i.e., from $178.99\ \mu J$ to $7.66\ \mu J$). Since the device exercises buffering time to buffer the rapid packets in a delivery, the effect of short inter packet arrival intervals is relieved. Therefore, as $V[t_a]$ increases, long packet arrival intervals are more significant. This is the reason that why $E_T^{(4)}$ decreases as $V[t_a]$ increases.

When λ_A is large (e.g., $\lambda_A = 50\lambda_R$), the packet arrival rate increases, which dominates the energy consumption on a device. Therefore, as λ_A increases, $E_T^{(4)}$ increases as well.

1.5.3 Effects of Variance of Inter REFRESH Interval

In this subsection, we analyze the delivery delay and energy consumption against the variance of inter REFRESH interval. For illustration purpose, we only observe these two output metrics of $M(4)$ (i.e., the device in the middle of the path). We assume that t_r has Gamma distribution with mean $1/\lambda_R$ and variance $V[t_r]$. Figure 1.7(a) depicts that delivery delay $E[t_s]$ is sensitive to the variance of inter REFRESH interval $V[t_r]$, especially when $V[t_r] > 1/\lambda_R^2$. As $V[t_r]$ increases, $E[t_s]$ increases as well. When $E[T] = 0.02/\lambda_R$, compared to $V[t_r] = 0.01/\lambda_R^2$, when $V[t_r] = 10/\lambda_R^2$, $E[t_s]$ increases from $0.74/\lambda_R$ to $5.75/\lambda_R$. Furthermore, $E[t_s]$ is not sensitive to the mean of buffering time $E[T]$ as well as $V[t_r]$. The maximum difference of delivery delay between $E[T] = 0.02/\lambda_R$ and $E[T] = 0.06/\lambda_R$ is about 22%, which is occurred when $V[t_r] = 0.001$. As $V[t_r]$ increases, the difference between $E[T] = 0.02/\lambda_R$ and $E[T] = 0.06/\lambda_R$ decreases.

Figure 1.7(b) illustrates the energy consumption against $V[t_r]$. It is significant that the energy consumption is unrelated to the configuration of REFRESH operation (e.g., mean and variance of inter REFRESH interval). Energy consumption is mainly related to mean buffering time. Compared to $E[T] = 0.02/\lambda_R$, $E[T] = 0.06/\lambda_R$ results in a significant decrement on energy consumption, where $E_T^{(4)}$ decreases from $161.73\mu J$ to $67.66\mu J$. Based on the observation from Figure 1.7, if devices are located in an environ-

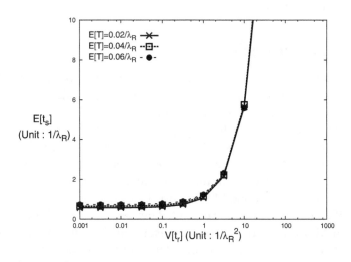

(a) $E[t_s]$

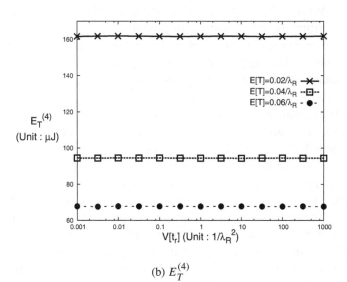

(b) $E_T^{(4)}$

Figure 1.7 Effects of $V[t_r]$

ment which has high variance of inter REFRESH interval, the devices should implement high mean buffering time (e.g., $0.06/\lambda_R$) to reduce the energy consumption of devices.

1.6 Open Challenges

Although data aggregation has been investigated for several years, there are still lots of challenges in the development of data aggregation. We choose two of the challenges to discuss in this section.

The first challenge is the design of routing algorithm and timing strategies. Most of the works of data aggregation is designed for wireless sensor network. However, wireless sensor network is likely to be replaced by the M2M network in the recent future. There are several differences between wireless sensor network and M2M network, such as the evaluation metric of energy efficiency, which we have pointed out in section 1.2. Furthermore, the devices in M2M network is more sophisticated than the sensor in wireless sensor network. We can apply more intelligent timing strategies, routing algorithms and aggregation mechanisms in these devices. For example, a high mobility environment results in significant costs in data aggregation routing algorithm. With Global Positioning System in M2M devices, we can propose a new algorithm based on the position and velocity of mobile devices to alleviate the problem of high mobility. Therefore, with sophisticated devices, the limitation of data aggregation is further released in M2M networks.

The second challenge is the security issues in data aggregation. The security issues of data aggregation are composed by two parts: (1) Aggregation Phase, (2) Transmission Phase. The characteristics of data aggregation result in the first security issue in data aggregation. In data aggregation, each packet needs to be decoded for aggregation. However, if there is a packet which carries some confidential information, it will be restricted to decode the packet. There is a trade-off between secrecy and additional energy consumption. The second security issue in data aggregation is resulted by the deployed environment of devices. When devices are deployed in a hostile environment, there are various security threats in transmission. As the complexity of encryption in transmission increases, the delivery delay and energy consumption will increase as well. Furthermore, if these devices are deployed in hostile environment, it is hard to provide direct power supply. The problem of security threats and energy consumption will be more severe. Therefore, we need to develop a new data aggregation technic which can support security transmission and various degrees of secrecy.

1.7 Conclusions

In this chapter, we investigated the timing strategy in data aggregation. Simulation results show that a slight increment on buffering time (e.g., $1/50\lambda_R$) results in a significant decrement in energy consumption on a packet (e.g., 88%). We also compare the output metrics in different environment (e.g., different inter packet arrival interval and inter REFRESH interval). We determine how to set the buffering timer in timing strategy to further improve the efficiency of data aggregation. As a final remark, the investment of timing strategy can combine with previous works that focused on issues like routing algorithms to improve the efficiency of M2M network.

References

[1] Whai-En Chen, Yi-Bing Lin, and Ren-Huang Liou. A weakly consistent scheme for ims presence service. *IEEE Transactions on Wireless Communications*, 8(7):3815–3821, Jul. 2009.

[2] Elena Fasolo, Michele Rossi, Joerg Widmer, and Michele Zorzi. In-network aggregation techniques for wireless sensor networks: A survey. *IEEE Wireless Communications*, 14(2):70–87, Apr. 2007.

[3] Matthias Handy, Marc Haase, and Dirk Timmermann. Low energy adaptive clustering hierarchy with deterministic cluster-head selection. In *Proc. of the 4th International Workshop on Mobile and Wireless Communications Network*, 2002.

[4] Wendi B. Heinzelman, Anantha P. Chandrakasan, and Hari Balakrishnan. Energy-efficient communication protocol for wireless microsensor networks. In *Proc. of the 33rd Annual Hawaii International Conference on System Sciences*, Jan. 2000.

[5] Wendi B. Heinzelman, Anantha P. Chandrakasan, and Hari Balakrishnan. An application-specific protocol architecture for wireless microsensor networks. *IEEE Transactions on Wireless Communications*, 1(4):660–670, Oct. 2002.

[6] Konstantinos Kalpakis, Koustuv Dasgupta, and Parag Namjoshi. Maximum lifetime data gathering and aggregation in wireless sensor networks. In *Proc. of the Joint International Conference on Wireless LANs and Home Networks*, 2002.

[7] Frank Kelly. *Reversibility and Stochastic Networks*. John Wiley & Sons, 1979.

[8] Samuel Madden, Michael J. Franklin, Joseph M. Hellerstein, and Wei Hong. Tag: a tiny aggregation service for ad-hoc sensor networks. *SIGOPS Operating Systems Review*, 36:131–146, Dec. 2002.

[9] Luca Mottola and Gian P. Picco. Muster: Adaptive energy-aware multisink routing in wireless sensor networks. *IEEE Transactions on Mobile Computing*, 10(12):1694–1709, Dec. 2011.

[10] Ramesh Rajagopalan and Pramod K. Varshney. Data-aggregation techniques in sensor networks: A survey. *IEEE Communications Surveys Tutorials*, 8(4):48–63, 2006.

[11] Ignacio Solis and Katia Obraczka. The impact of timing in data aggregation for sensor networks. In *IEEE International Conference on Communications*, Jun. 2004.

[12] Sok-Ian Sou and Chuan-Sheng Lin. Spr proxy mechanism for 3gpp policy and charging control system. *Computer Network*, 55(17):3847–3862, Dec. 2011.

[13] Meng-Hsun Tsai and Hui-Wen Dai. Bearer reservation with preemption for voice call continuity. In *Proc. of International Conference on Mobile Data Management: Systems, Services and Middleware*, May 2009.

[14] A. Wood, G. Virone, T. Doan, Q. Cao, L. Selavo, Y. Wu, L. Fang, Z. He, S. Lin, and J. Stankovic. Alarm-net: Wireless sensor networks for assisted-living and residential monitoring. Technical report, 2006.

[15] Yong Yao and Johannes Gehrke. The cougar approach to in-network query processing in sensor networks. *SIGMOD Rec.*, 31(3):9–18, Sep. 2002.

2

OR-AHaD: An Opportunistic Routing Algorithm for Energy Harvesting WSN

Sanam Shirazi Beheshtiha[1], Hwee-Pink Tan[2] and Masoud Sabaei[1]

[1]Computer Engineering and Information Technology Department, Amirkabir University of Technology, Tehran, Iran
[2]Sense and Sense-abilities Program, Institute for Infocomm Research (I2R), A*STAR, Singapore
e-mail: {s.shirazi@aut.ac.ir, hptan@i2r.a-star.edu.sg and sabaei@aut.ac.ir}

Abstract

With recent advances, the trend has shifted from battery-powered wireless sensor networks towards ones powered by ambient energy harvesters (WSN-HEAP). In such networks, operability of the node is dependent on the harvesting rate which is usually stochastic in nature. Therefore, it is necessary to devise routing protocols with energy management capabilities that consider variations in the availability of the environmental energy. In this book chapter, we design OR-AHaD, an Opportunistic Routing algorithm with Adaptive Harvesting-aware Duty Cycling. In the proposed algorithm, candidates are primarily prioritized by applying geographical zoning and later coordinated in a timer-based fashion by exchanging coordination messages. An energy management model is presented which uses the estimated harvesting rate in the near future to adjust the duty cycle of each node adaptively. Simulation results show that OR-AHaD exploits the available energy resources in an efficient way and increases goodput in comparison to other opportunistic routing protocols for WSN-HEAP.

Keywords: opportunistic routing, harvesting-aware, wireless sensor networks.

Fabrice Theoleyre and Ai-Chun Pang (Eds.), Internet of Things and M2M Communications, 23–47.

2.1 Introduction

Energy limitation is a major challenge which has been addressed many times in the context of wireless sensor networks. Traditionally, sensor nodes were equipped with batteries as their main source of power and their lifetime was limited to their battery life. However, recently with the emergence of energy harvesting technologies for embedded systems, the trend has headed towards wireless sensor networks which are powered by ambient energy harvesters (WSN-HEAP) with the advantages of being cheaper, easier to deploy, more environmentally friendly and most importantly capable of being recharged many times [17].

Different types of energies such as solar, thermal and vibrational can be exploited using a variety of energy harvesting devices. Availability of the energy depends on many factors including the internal structure of the harvesting device as well as temporal conditions and the location of deployment. The energy harvesting capability of current technologies starts from tens of μW but does not exceed several mW. For instance, a 10 cm^2 electromagnetic vibrational device can harvest on average 40 μW, whereas a typical indoor solar device of the same size can harvest up to but no more than 37 mW [17]. The typical power consumption of a node's transceiver during its operation is much more than the average replenishment rate (around 3 to 20 times) which is usually stochastic in nature. Consequently, the energy harvesting sensor node cannot remain active continuously and needs to shutdown from time to time. Based on these energy characteristics, it is important to develop novel energy management models and routing protocols that consider the temporal and spatial variations of the available environmental energy.

Several routing strategies have been suggested for WSN-HEAP. Until recently, most of them focused on traditional routing techniques which are not very suitable for such networks considering the stochastic variations in harvesting rates. In order to overcome, opportunistic routing has been suggested to best exploit the available environmental energy and increase goodput by relying on the broadcast nature of the wireless medium.

However, the existing proposal on opportunistic routing for WSN-HEAP has some limitations [4]. Firstly, the energy management model determines the duty cycle of each node independently of the future availability and assigns schedules only based on the average and current harvesting rate. Moreover, coordination among the candidates for the next hop is time consuming and therefore, goodput is not as high as it can be.

In this book chapter, we present and evaluate an Opportunistic Routing algorithm with Adaptive Harvesting-aware Duty Cycling (OR-AHaD). The key contributions are: First, a coordination message is used instead of the original data packet to shorten the overall coordination interval. Second, the energy model is modified to incorporate the exchange of coordination messages. Based on that, an energy management model is proposed that adaptively adjusts the duty cycle by considering not only the current, but also the near future availability of energy. Third, the number of geographical zones, is recomputed based on the new energy model.

The remainder of this chapter is organized as follows. The related work is presented in Section 2.2. Next, Section 2.3 presents the network conditions and energy model under which the proposed algorithm is evaluated. Followed by that, Section 2.4 describes OR-AHaD algorithm and its features in detail. The performance evaluation results using simulation are discussed in Section 2.5. Then in Section 2.6, the remaining open challenges in this area are argued. Lastly, conclusions are drawn in Section 2.7.

2.2 Related Work

In the literature, a number of routing solutions have been proposed for WSN-HEAP. They are mostly traditional techniques aimed for increasing efficiency by optimizing energy consumption or goodput by considering geographical locations. On the other hand, recently the idea of opportunistic routing has been brought up in this context. It promotes taking advantage of the broadcasting nature of wireless channels and making per-hop decisions after packet reception. In this section, we will have a brief look at the existing work.

2.2.1 Traditional Energy-Aware and Geographical Routing

The most common approach towards routing in wireless networks is the conventional routing where the routing path is determined before or upon transmission of the packet based on a cost metric reflecting the routing goals. WSN-HEAP is no exception to that with the main attention towards geographical distances and energy efficiency. Some have defined routing as an energy efficiency problem and worked on choosing energy efficient routes or integrating energy management functions, while others involved geographical advancement.

Some research studies have been carried out on power management of WSN-HEAP (eg. [8, 12, 14]). They mostly rely on designing low overhead algorithms for the prediction of the future harvesting rate. In [14], a method is proposed for predicting the future availability of energy based on the current energy sample and samples of previous days for solar harvesters meanwhile taking into account the changes in weather conditions. Another study extends power management for solar harvesters to real-time applications using time series and DVFS techniques. Regression analysis, moving average and exponential smoothing are the three methods suggested for run-time prediction in [12]. Having a time-stamped value of sunlight intensity, a linear regression model can predict the future harvesting rate. Moving average is another method that forecasts based on the average value of the past observations. Exponential smoothing operates the same way as moving average, only with weights relative to the freshness of the past observations.

Several energy-aware routing solutions have been proposed for WSN-HEAP. The authors of [9] described the energy efficiency problem as achieving maximum workload under available energy rather than increasing lifetime which was tailored to battery-powered sensors and sought for an optimal routing solution based on that. The probability of forwarding the packet on each link is proportional to the the maximum flow through that link. In another work, a mathematical framework was developed to parameterize the real characteristics of environmental energy. Then analytical strategies were associated with it to evaluate the benefits of energy-aware routing in the presence of renewable energy sources [11]. A cost value is computed for every node based on the battery capacity, available energy, harvesting and consumption power and then the shortest path is calculated accordingly.

While a few of the existing studies explicitly consider the time varying environmental energy such as the above-mentioned, many of them still concentrate only on the residual energy level. For instance, a modified version of AODV is designed which uses the weighted value of the residual energy and hop count as the metric for battery powered sensor networks with complementary solar harvesters to increase the battery lifetime [3].

Some other studies have focused on merging geographical routing with energy-awareness. Geographical routing with the benefit of low overhead, scalability and high capacity has shown to be a proper choice for many WSN applications that require location awareness. In geographical routing, the decision on data progress is based on the location information of the node, its neighbors and the sink. Having this information, data can be directed to a particular region and progress towards the sink at each hop. In [19], the energy model of a solar harvester is incorporated into geographic routing. Nodes are assumed to periodically exchange their current residual energy and expected harvesting rate with their neighbors. Based on the gathered information, each node forwards the received packet to the neighbor that minimizes the cost based on the progressive distance and effective energy. In [13], the dissemination scope of topological information is adjusted adaptively based on the solar energy budget over the next period and packets are routed according to their QoS constraints related to delay sensitivity. A distributed routing scheme is presented which aims for the energy optimized routes [7]. Initially, this routing scheme finds all the shortest paths which are also the ones with the least energy consumption. Then it maps the available energy to a local distance penalty on each path and solves the local minimum problem by a distributed penalty metric. The final shortest path is recomputed considering these penalties and the stochastic nature of harvesting rate is considered through some global parameters.

Traditional routing does not reach its full potential in WSN-HEAP because the stochastic variation in environmental energy causes uncertainty about the near future and results in asynchronous schedules. Therefore the set of potential next-hop nodes cannot be known at a reasonable cost prior to sending the packet. In this case, opportunistic routing seems to be a proper solution.

2.2.2 Opportunistic Routing and EHOR

In order to overcome the limitations of traditional routing algorithms for WSN-HEAP, opportunistic routing has been suggested. By benefiting from the nature of the wireless channel, a packet is broadcasted at each step and then any decisions on the next hop selection is deferred until the successful reception by the available neighbors. After that the routing comprises: (a) filtering the potential candidates (b) priority designation to the filtered candidates and (c) coordinated transmission based on priority [6].

The idea of exploiting opportunistic geographic routing for WSN-HEAP was first proposed in [4]. To the best of our knowledge, this is the only work that has been done in this area up to now. It proposes EHOR, a novel routing technique that aims at improving the shortcomings of conventional opportunistic routing schemes suited for battery-powered WSN.

EHOR uses a regioning approach. Filtering is done based on geographical advancement. It means that from the nodes that have successfully received the packet, only the ones that are closer to the destination than the previous hop are eligible to be a candidate. Among the filtered candidates, priority is determined individually based on the distance to the sink and the residual energy using weighted averaging. After that, coordination takes place in a time-slotted manner where the higher priority nodes are assigned to earlier slots. In the assigned slot, the node transmits the data packet only if it has not overheard others' transmissions in the previous slots.

The performance of opportunistic routing in such networks also depends on the associated energy model and the schedule of the nodes. In EHOR, at the beginning of each cycle, each node remains inactive until the stored energy reaches a defined threshold and therefore, the duration depends on the current harvesting rate. After that, the node becomes active for a fixed amount of time which is the same in every cycle and is dependent on the average harvesting rate.

Two limitations of EHOR are:

- Future availability of the energy is not taken into consideration and the schedule of a node is only based on the current and average harvesting rate. Therefore, the energy is not managed optimally.
- Even though regioning is applied and coordinating delay is reduced, it is still not maximally reduced and therefore, affects the goodput.

The work presented in this book chapter is a follow up on EHOR and is aimed at addressing the above mentioned limitations.

2.3 Network Model and Assumptions

The network model must captures the energy model which consists of energy consumption and energy replenishment. It also describes the deployment topology and traffic properties of the network. The notations used in this paper are summarized in Table 2.1.

2.3.1 Topology and Traffic Characteristics

The network consists of $n_{sensor} = 20$–300 sensor nodes in a 1D area spanning $d_{max} = 300$ m. All of the sensor nodes play the role of forwarding relays in a multihop routing scenario but only one of them acts as the data source. This is realistic in event-triggered applications such as target tracking. The sink node is located at the origin of the coordinate system. $R = 70$ m is the maximum transmission range of the node where packet delivery ratio is above 10% as

Table 2.1 Notations used for describing OR-AHaD.

Symbol	Denotes
d_{max}	Length of the deployment area
d_{pre}	Candidate's distance from the previous hop
$E_{con-act}$	Energy consumed in an active period
$E_{con-act}$	Energy consumed in an active period
$E_{har-act(i)}$	Energy harvested in active period i
E_{max}	Energy of a fully charged sensor node
$E_{res-act(i)}$	Residual Energy at the end of active period i
$E_{av(i)}$	Available energy for active period i
M_j	Sensor node's mode ($j = s$: sleep, $j = r$: receive, $j = t$: transmit, $j = c$: coordination)
n_{sensor}	Number of sensor nodes
N_{act}	Number of active periods in a cycle
N_{zones}	Number of geographical zones
p_{act}	Probability of being active and in the receive mode
$p_{end-act}$	Probability of ending a cycle and going to inactive mode
$P_{har-act(i)}$	Harvesting rate in active period i
$P_{har-avg}$	Average harvesting rate
P_{rx}	Receive power of the sensor node
P_{tx}	transmission power of the sensor node
r	Transmission rate of the sensor node
R	Maximum transmission range where packet delivery ratio is above 10%
s_c	Size of the coordination packet
s_d	Size of the data packet
s_{dist}	Slot number for candidate transmission based on the distance factor
s_{ene}	Slot number for candidate transmission based on the energy factor
s_{no}	Final slot number for candidate transmission
$slot_c$	Slot duration for sending a coordination packet
$slot_d$	Slot duration for sending a data packet
t_{act}	Duration of an active period
$t_{har-act(i)}$	Charge time in active period i
$t_{har-avg}$	Average charge time of a depleted node to level E_{max}
t_{inact}	Duration of an inactive period
t_{M_j}	Duration of mode M_j
t_{prop}	Maximum propagation delay
t_{tu}	Hardware turnaround delay from receive to transmit state and vice versa
γ	Candidate priority weight between distance and energy factors

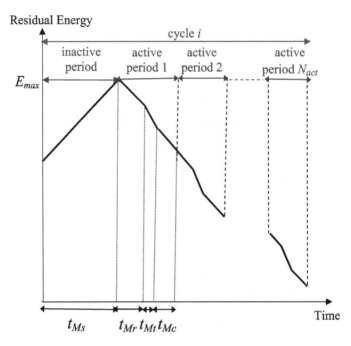

Figure 2.1 Energy model of a sensor node.

in [4]. We use Log-normal shadowing and Ricean fading to model the radio propagation. All the nodes are equipped with GPS.

We study a saturated network where the source either forwards a received data packet or sends a new data packet in each active period explained in Section 2.4.2. The data packet and the coordination message size are $s_d = 100$ bytes and $s_c = 15$ bytes, respectively. In this network, the channel rate is $r = 250$ Kbps. The propagation delay and the hardware turn-around time (from receive to transmit) are $t_{prop} = 0.008$ ms and $t_{tu} = 0.192$ ms, respectively.

2.3.2 Energy Model

Sensor nodes are powered by energy harvesters. The specification of TI energy harvesting sensor nodes are assumed in our scenarios [18]. Energy is harvested at all times. For the energy storage, a 12μAh Enerchip rechargeable battery is used with output voltage of 3.8 V [1]. Sink is connected to the power supply and does not require recharging.

The proposed energy model of a sensor node is as illustrated in Figure 2.1. Each node runs through a number of cycles. Each cycle consists of one inactive period, followed by a couple of active periods. The number of active periods in a cycle is determined by an adaptive harvesting-aware duty cycling algorithm which will be introduced later in this chapter.

The duration of inactive period t_{M_s}, depends on the harvesting rate which is different in each period. We assume that the nodes are aware of their harvesting rate in the current period and can also predict the rate of the next period. An exponential distribution is used to model the time it takes for a depleted node to get charged to a level denoted as E_{max}. The parameter of this exponential distribution is the inverted average charge time (computed using E_{max} and the given average harvesting rate of the scenario denoted as $P_{har-avg}$). If the charge time in the active period i is $t_{har-act(i)}$, then the harvesting rate in that period can be derived from

$$P_{har-act(i)} = \frac{E_{max}}{t_{har-act(i)}}.$$

In the inactive period, the node remains in sleep mode M_s until the battery is charged to E_{max}. The power consumption in this mode is negligible since the node's components are mostly shut down. The duration of the inactive period t_{M_s}, depends on the harvesting rate in that period.

The active period initiates in receive mode M_r with duration t_{M_r}. The node waits to receive a data packet. The power consumption in this mode is $P_{rx} = 72.6$ mW. After that it will shift to transmit mode M_t to send coordination message and forward the received packet. The interval of this mode is t_{M_t} and the power consumed is $P_{tx} = 83.7$ mW. Once the transmit mode is over, the node transits to coordination mode M_c, where it waits for the specified time t_{M_c} to receive a coordination message for the packet forwarded earlier to make sure it has progressed. The power usage is the same as receive mode.

We let the $min(t_{M_r})$, $max(t_{M_r})$ and $E[t_{M_r}]$ be the minimum, maximum and expected time in the receive mode, respectively. Now the minimum, maximum and expected energy consumption of a node in each active period is computed as:

$$min(E_{con-act}) = min(t_{M_r}) \cdot P_{rx} + t_{M_t} \cdot P_{tx} + t_{M_c} \cdot P_{rx} \qquad (2.1)$$

$$max(E_{con-act}) = max(t_{M_r}) \cdot P_{rx} + t_{M_t} \cdot P_{tx} + t_{M_c} \cdot P_{rx} \qquad (2.2)$$

$$E[E_{con-act}] = E[t_{M_r}] \cdot P_{rx} + t_{M_t} \cdot P_{tx} + t_{M_c} \cdot P_{rx} \qquad (2.3)$$

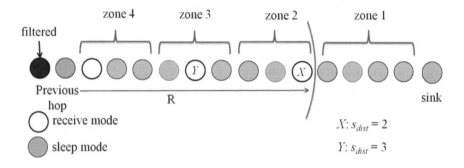

Figure 2.2 Transmission priority based on the distance factor in a 1D area with 4 zones.

The amount energy harvested in active period i is:

$$E_{har-act(i)} = \left(t_{M_r} + t_{M_t} + t_{M_c}\right) \cdot P_{har-act(i)} \tag{2.4}$$

2.4 Opportunistic Routing with Adaptive Harvesting-aware Duty Cycling (OR-AHaD)

To describe OR-AhaD in details, first different phases of routing in the proposed algorithm are explained. Then zoning and coordination messages and their application are discussed. Lastly the energy management model which adaptively adjusts the duty cycle of the nodes is presented.

2.4.1 Routing Algorithm

As we said earlier in Section 2.2.1, any opportunistic routing protocol is composed of three phases. In the initial phase of OR-AHaD, the candidates who have received the packet but are not closer to the sink than the previous hop are filtered. In the subsequent step, priority designation is done via geographical zone assignment. Nodes residing in the zones closer to the sink have higher transmission priorities. Afterwards, a timer-slotted coordination strategy is applied and the candidates are delayed according to their priorities. The number of slots is equal to the number of zones and the duration of each slot is fitted for sending a short coordination message. Detailed information regarding zones can be found in Section 2.4.2.

Transmission priorities can be computed as in [4]. We let the distance from the previous hop to the current candidate be d_{pre} and the number of zones as N_{zones}. The slot number for transmitting the coordination message based on the distance factor, s_{dist}, is the same as the zone number. Zones closer to the destination have higher priorities and consequently are assigned earlier slots as illustrated in the example in Figure 2.2. Each candidate can compute s_{dist} using

$$s_{dist} = \begin{cases} 1 + \lceil (1 - \frac{d_{pre}}{R}) \cdot (N_{zones} - 1) \rceil, & d_{pre} \leq R \\ 1, & R < d_{pre} \end{cases} \quad (2.5)$$

The performance can be improved by improvising a scheme to adjust the final transmission priority based on the residual energy, in addition to distance from the sender as in [4]. Accordingly if the residual level of the node after receiving the packets is E_{res}, then the final transmission slot denoted as s_{no} can be calculated using

$$s_{no} = \gamma \cdot s_{dist} + (1 - \gamma) \cdot s_{ene} \quad (2.6)$$

where γ is a weighting factor and s_{ene} is calculated using:

$$s_{ene} = \left\lceil \frac{E_{res}}{E_{max}} \cdot N_{zones} \right\rceil \quad (2.7)$$

2.4.2 Zoning and Coordination Messages

We apply the concept of geographical zoning as in [4], for two main reasons. The first one is reducing the coordination delay which is a major challenge in timer-based coordination methods. Instead of assigning slots per node, we designate it per zone. Therefore, the overall interval is shortened. The second advantage is enabling nodes to individually determine their priorities with no extra knowledge required from the other candidates.

There is a trade-off between the end-to-end delay and the the probability of collision in determining the number of zones [4]. N_{zones} must be computed in such a way that one and only one active node resides in it. We use approximations in computing the number of zones. We dedicate one zone to the candidates outside the transmission range of the source who still might receive the data packet with low probability. As for the rest, assuming that nodes are uniformly distributed across the deployment area, the number of candidates who are closer to sink than previous hop and fall within the transmission range would be $n_{sensor} \cdot \frac{R}{d_{max}}$.

Assuming each node is in active period with probability $prob_{act}$, then N_{zones} must be equal to the number of active nodes to serve our purpose and hence, it can be obtained by solving

$$N_{zones} = \left\lceil prob_{act} \cdot n_{sensor} \cdot \frac{R}{d_{max}} \right\rceil + 1 \qquad (2.8)$$

This equation is similar to the one for computing the number of regions in [4]. However, $prob_{act}$ must be computed differently because the energy model and schedule of the nodes are different. Before deriving an equation for $prob_{act}$, we first have to describe the application of coordination messages and compute the time the node spends in active and inactive periods.

Upon reception of a packet in mode M_r, the filtered candidate computes its priority and start its coordination timer. From the start of the interval, it keeps listening to the channel to see if any active candidate with higher priority declares to be the next official hop. Once a node overhears declaration in earlier slots, it abstains from forwarding and resets its coordination timer and continues listening for other incoming data packets until the maximum receive time is over. In the proposed algorithm instead of sending out the data packet itself, special coordination messages are used. In this way each slot can be fixed for transmission of a short $s_c = 15$ bytes coordination message at MAC layer instead of the data packet which is $s_d = 100$ bytes. As a result the overall coordination time is reduced even more. According to this, we have two kinds of slots. One for transmitting data packets denoted as

$$slot_d = t_{prop} + \frac{s_d}{r} + t_{tu}$$

and the other for transmitting coordination messages specified as

$$slot_c = t_{prop} + \frac{s_c}{r} + t_{tu}.$$

Mode M_r ends either when it is time for a node to forward a received packet or when the timer indicating the maximum time goes off. In this mode, it takes as long as $slot_d$ to fully receive a data packet. Afterwards, the node computes its transmission slot. In the best case it is designated the first slot so it immediately transfers to M_t. Therefore, the minimum time in the receive mode is:

$$min(t_{M_r}) = slot_d \qquad (2.9)$$

However, in the worst case, it is in the zone with the lowest priority and hence has to stay in this mode during the whole coordination interval. The maximum time that a node is allowed to stay in M_r, should be greater than the entire coordination interval in addition to $slot_d$. In this book chapter, we let the maximum time in the receive mode be:

$$max(t_{M_r}) = slot_d + N_{zones} \cdot slot_c \qquad (2.10)$$

The expected value for t_{M_t} can be computed based on 2.9 and 2.10:

$$E[t_{M_r}] = \frac{min(t_{M_r}) + max(t_{M_r})}{2} \qquad (2.11)$$

The winner immediately proceeds to mode M_t. If it is only a relay node, it sends a coordination message followed by the data packet which is due for transmission. However, if it is also a data source and does not have a packet to forward at that time, it transmits a new data packet. The node uses carrier sensing before sending a packet to make sure the channel is idle. Transmission of coordination and data packets take place in this mode with the fixed duration t_{M_t}:

$$t_{M_t} = slot_d + slot_c \qquad (2.12)$$

Having transmitted the packet, the node transits to mode M_c and waits for the reception of the coordination message from the next hop candidate. This process takes as long as the coordination interval:

$$t_{M_c} = N_{zones} \cdot slot_c \qquad (2.13)$$

According to the above equations, the expected value for the total time in active period denoted as t_{act} is:

$$E[t_{act}] = E[t_{M_t}] + t_{M_t} + t_{M_c} \qquad (2.14)$$

The Zoning Approach degrades the possibility of concurrent transmissions in a slot. However, in the case of collision, candidates in subsequent slots have the chance of forwarding the data packet. The probability of duplicate transmission is reduced by listening to the channel while others are sending out coordination messages, but cannot be avoided for the candidates outside the overhearing range.

After going through a number of active periods, N_{act}, which is based on the output of the adaptive algorithm in Section 2.4.3, node finally goes to inactive mode and remains there until it is fully charged. The amount of energy that needs to be replenished depends on the number of active periods and the energy left at the end of each period. The expected value of t_{M_s} (same as t_{inact}) is calculated from:

$$E[t_{M_s}] = \frac{E_{max} - E[N_{act}](E[E_{con-act}] - P_{har-avg} \cdot E[t_{act}])}{P_{har-avg}} \tag{2.15}$$

With this information the probability of being in active mode for a node in (2.8) can finally be computed as the ratio of the average time in the receive mode to the average duration of a cycle:

$$prob_{act} = \frac{E[N_{act}] \cdot E[t_{M_r}]}{E[N_{act}] \cdot E[t_{act}] + E[t_{inact}]} \tag{2.16}$$

The duration of the receive mode and the whole cycle depends on $E[N_{act}]$, the average number of active periods in a cycle. Rearranging (2.8) and (2.16), the value of N_{zones} can be obtained by solving a quadratic equation in $O(1)$ time, which always yields a nonnegative result.

2.4.3 Adaptive Harvesting-Aware Duty Cycle Management

In a battery-powered WSN for the purpose of energy conservation and increase in network lifetime, it is desirable that nodes spend most of their time in the sleep mode as long as they meet the QoS constraints. However, in WSN-HEAP the challenge is exploiting the available environmental energy to meet the required QoS and hence it is important to adapt to the changing environment. Here we propose a harvesting-aware energy management model which determines node's schedule adaptively.

We presented the energy model of the node in Section 2.3.2. As we said each sensor node runs through a number of cycles which is composed of an inactive period, followed by a number of active periods. Instead of using a fixed number of active periods in each cycle throughout the scenario, we use the knowledge of the current and near future energy availability to decide at the end of each period whether to end the cycle and transit to sleep mode or continue the cycle by going to the next active period.

We use three energy factors as input to the decision making model: (a) the residual energy at the end of the current period (b) the expected energy harvested during the next period which is computed using (2.4) based on the predicted value of $P_{har-act}$ and (c) the minimum and maximum energy consumption in a single active period which can be derived from (2.1) and (2.2), respectively.

We Assume at the end of active period $i - 1$ of a cycle, the residual energy is $E_{res-act(i-1)}$ and the expected energy harvested during the next active

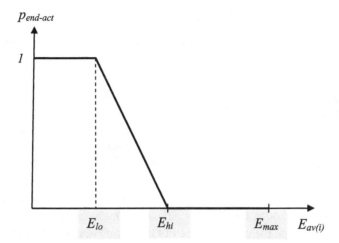

Figure 2.3 Probability density function of ending the cycle after active period $i - 1$.

period, if we decide to proceed, would be $E[E_{har-act(i)}]$. Consequently, the available energy for period i can be computed as:

$$E_{av(i)} = E_{res-act(i-1)} + E[E_{har-act(i)}]$$ (2.17)

We define two thresholds E_{lo} and E_{hi} as the minimum and maximum energy consumption in each active period:

$$E_{lo} = min(E_{con-act})$$ (2.18)

$$E_{hi} = max(E_{con-act})$$ (2.19)

Then we use probabilistic decision making. The probability of going to the next cycle based on environmental energy factors is $p_{end-act}$ and the probability density function is illustrated in Figure 2.3 and formulated in (2.20). According to that, if $E_{av(i)}$ is less than E_{lo}, node definitely ends the cycle and goes to inactive period. If it is more than E_{hi}, it definitely proceeds to active period i. Otherwise, the probability is decreased linearly.

$$P_{end-act} = \begin{cases} 1, & E_{av(i)} < E_{lo} \\ \frac{E_{hi}-E_{av(i)}}{E_{hi}-E_{lo}}, & E_{lo} \leq E_{av(i)} < E_{hi} \\ 0 & E_{hi} \leq E_{av(i)} \end{cases}$$ (2.20)

2.5 Performance Evaluation

We use the QualNet Network Simulator [2] to evaluate the performance of OR-AHaD. We implement the proposed energy model and the routing algorithm into this simulator. The performance metrics used for evaluation are: (a) *Goodput*: rate of receiving non-duplicate data packets at sink (b) *Efficiency*: ratio of non-duplicate data packets to all the data packets received at sink (c) *Data Delivery Ratio*: ratio of the non-duplicate data packets received at sink to the data packets sent by the source and (d) *Hopcount*: the average number of hops traversed until a data packet is received at sink.

The network model including deployment and traffic characteristics are as explained in Section 2.3.1. In all scenarios, nodes are assumed to be TI energy harvesting sensor nodes with the specifications mentioned in Section 2.3.2.

We first evaluate the performance of OR-AHaD under different scenarios. Then we compare its performance with EHOR. Data results presented in all the scenarios are derived by averaging 20 simulation runs using different seeds. The simulation time is 200 s with the warm-up period of 5 s.

2.5.1 Effect of Varying γ

The key parameter in the design of OR-AHaD is γ, $0 \leq \gamma \leq 1$, a factor that weights the transmission priority of a candidate between its distance to the sink and its residual energy. When $\gamma = 0$, candidates with lower residual energy are designated with higher transmission priorities. Here we study the effect of different values of γ on the overall performance of OR-AHaD. We set $P_{har-avg} = 10$ mW.

We first assume a scenario where all the sensor nodes including the source are randomly deployed. The simulation results are as illustrated in Figure 2.4. In another scenario, the source node is deployed such that it is furthest way from the sink while the other sensor nodes are still deployed in a random way and the simulations are repeated. The performance results are presented in Figure 2.5. The overall performance in the first scenario is better than the second one because multi-hop routing is more challenging when the distance between source and sink is high.

In both scenarios, increasing γ improves goodput and delivery ratio. By giving more weight to the distance factor, advancement per hop increases so packets suffer less delay and the network does not get too congested. The limitation of energy will not pose much of a problem because the energy management model already takes care of this and manages the candidate set adaptively according to environmental energy factors. Hopcount also de-

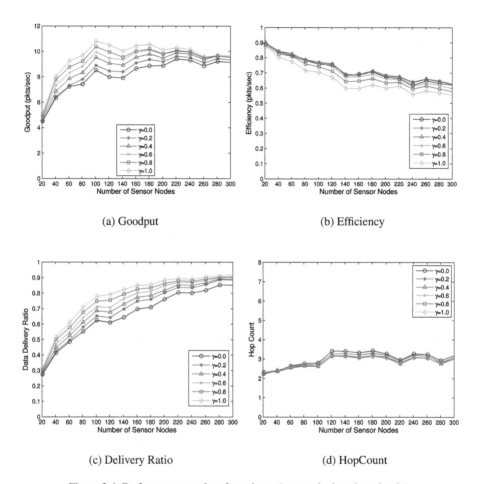

(a) Goodput

(b) Efficiency

(c) Delivery Ratio

(d) HopCount

Figure 2.4 Performance results of varying γ (source deployed randomly).

creases because fewer distant candidates from sink are assigned with higher priorities. However as γ increases, the efficiency drops. When $\gamma = 1.0$, the nodes outside the transmission rage of the source are designated with the highest priority. When they use coordination messages to inform other candidates, many of them are outside the overhearing zone and retransmit the packet in the following slots. As a result, the rate of receiving duplicate packets at the sink, increases and efficiency drops.

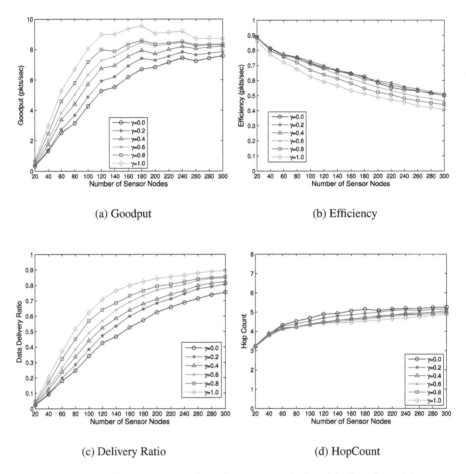

(a) Goodput

(b) Efficiency

(c) Delivery Ratio

(d) HopCount

Figure 2.5 Performance results of varying γ (source deployed furthest from sink).

2.5.2 Effect of Varying Average Harvesting Rate

In this scenario we study the effects of changing the average harvesting rate which symbolizes a wide range of energy harvesting devices and various temporal and spatial conditions. Figure 2.6 illustrates the scenario with 300 sensor nodes, $P_{har-avg} = 3–27$ mW. γ is assumed it to be 1.0 which gives the highest goodput. Sensor nodes including the source are randomly deployed.

As the rate increases, nodes spend most of their time in active mode rather than inactive mode. The multiplicity of zones grows because of the increase in the number of nodes that are not asleep. This extends the candidate set

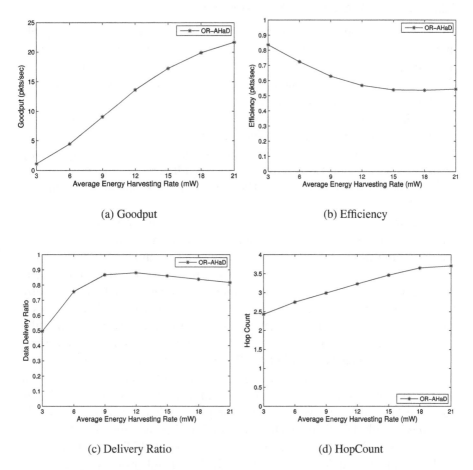

(a) Goodput

(b) Efficiency

(c) Delivery Ratio

(d) HopCount

Figure 2.6 Performance results of varying average harvesting rates.

and as a result enhances goodput and delivery ratio. However, after some point, delivery ratio decreases slightly because of an excessive number of candidates. As for efficiency, it decreases because the increment in number of zones, scales-up the likelihood of duplicate reception. The increase in the likelihood of multi-path affects average hop count as well.

2.5.3 Comparing OR-AHaD with EHOR

To assess the performance gain of using adaptive harvesting-aware duty cycle management and combining coordination messages with geographical zoning, we compare our algorithm with EHOR. We consider different values of β which is the weighting factor for the transmission priority in EHOR [4]. We assume γ to be 1.0 in our algorithm for achieving the highest goodput.

We first assume a scenario where all the sensor nodes including the source are randomly deployed. The simulation results are as illustrated in Figure 2.7. In another scenario, the source node is deployed such that it is furthest way from the sink while the other sensor nodes are still deployed in a random way and the simulations are repeated. The performance results are presented in Figure 2.8.

In both scenarios, OR-AHaD achieves much higher goodput in comparison to EHOR. Also, since the energy management model is based on the prediction of the future harvesting rate, a more suitable candidate set is available and active to contribute to forwarding the packets. The efficiency is also higher in OR-AHaD because the rate of duplicate packets is limited by using coordination messages. Even at the last hop, when the sink receives a data packet, it sends out coordination message like previous hops along the path. This way, all of the neighbor nodes are informed and prevented from relaying that data packet. But in EHOR when sink receives the data packet, it does not notify the neighbor nodes. This can affect efficiency especially when the number of nodes is relatively high. With delivery ratio, the reasoning is the same as goodput. However, after some point, delivery ratio in OR-AHaD becomes slightly less than EHOR . As mentioned before, the rate of sending new packets in OR-AHaD is higher in comparison to EHOR. Therefore, when we have a large candidate set because of the high number of deployed nodes, network load caused by multi-hop routing increases even more and hence, some packets may not make it to the sink. Hop count is almost the same in both algorithms.

2.6 Open Challenges

While a lot of efforts have been invested towards the design and adaptation of novel routing solutions for wireless sensor networks that take advantage of environmental energy to either supplement batteries or as the main source of power in the last few years, this research area is still relatively new, with an

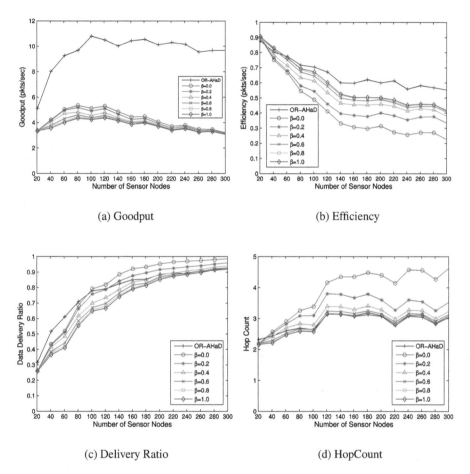

(a) Goodput

(b) Efficiency

(c) Delivery Ratio

(d) HopCount

Figure 2.7 Comparing performance results of OR-AHaD with EHOR (source deployed randomly).

abundance of challenges that need to be investigated further. We dedicate this section to discussing some of these open challenges.

The core part of an energy harvesting sensor node is the harvesting system that scavenges energy from the environment [17]. The uncertainty and instability in energy availability that is inherent in ambient energy sources intensifies the role of harvesting models and prediction methods in the performance of routing algorithms. Therefore, it is essential to model energy harvesting and parametize the temporal, spatial and internal characteristics

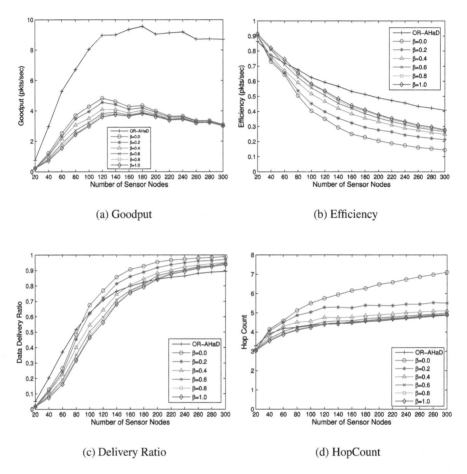

(a) Goodput

(b) Efficiency

(c) Delivery Ratio

(d) HopCount

Figure 2.8 Comparing performance results of OR-AHaD with EHOR (source deployed furthest from sink).

of harvesting devices, so that this can be exploited in routing decisions. On top of that, to increase the accuracy of forecasting, customized methods must be adopted for the specific type of harvesting device that is being utilized. While most of the existing methods concentrate on solar energy harvesting devices, less attention has been paid towards other types including thermal and vibrational (eg. [12, 14]). In addition, the effect of prediction accuracy in the performance of routing must be investigated extensively.

Some research activities have been conducted to evaluate routing algorithms in WSN-HEAP [5]. However, a unified evaluation framework is required to compare the existing solutions based on realistic MAC and physical layer specifications. This would give a heads up on the direction to be taken for improving the performance of routing in WSN-HEAP. Theoretical bounds need to be derived for critical performance metrics such as goodput and delay to understand where the current solutions stand and how far they can progress. In addition to theoretical and simulation studies, existing routing algorithms should be testbedded to effectively take into consideration implementation issues, and to help realize the potential of WSN-HEAP in actual deployments.

Finally, possible security threats imposed by nature of the WSN-HEAP must be identified, from which preventive measures and solutions can be investigated and integrated with future routing and data dissemination solutions.

2.7 Conclusion

In this book chapter, we presented OR-AHaD, an Opportunistic Routing Algorithm with Adaptive Harvesting-aware Duty Cycling for WSN-HEAP. We proposed an energy management model which exploits the estimated value of harvesting rate in the near future and the residual energy to adjust the duty cycle of each node adaptively and integrated that in the opportunistic routing algorithm which prioritizes the candidates based on geographical information and by applying a zoning approach. Then, we introduced the use of coordination messages among candidates in a timer-based coordination method.

We evaluated OR-AHaD using extensive simulation. The results showed that goodput and efficiency is increased in comparison to EHOR, which is a previously proposed opportunistic routing scheme for WSN-HEAP. The proposed algorithm also performs well under different environmental energy harvesting rates. In the last section, we addressed the remaining open challenges in this area for future work.

References

[1] Cbc012 enerchip. http://www.cymbet.com/pdfs/DS-72-02.pdf.
[2] Qualnet network simulator. http://www.scalable-networks.com.

[3] Hung-Chi Chu, Wei-Tsung Siao, Wei-Tai Wu, and Sheng-Chih Huang. Design and implementation an energy-aware routing mechanism for solar wireless sensor networks. In *High Performance Computing and Communications (HPCC), 2011 IEEE 13th International Conference on*, pages 881–886, Sept. 2011.

[4] Zhi Ang Eu, Hwee-Pink Tan, and Winston K.G. Seah. Opportunistic routing in wireless sensor networks powered by ambient energy harvesting. *Computer Networks*, 54(17):2943–2966, 2010.

[5] David Hasenfratz, Andreas Meier, Clemens Moser, Jian-Jia Chen, and Lothar Thiele. Analysis, comparison, and optimization of routing protocols for energy harvesting wireless sensor networks. In *Sensor Networks, Ubiquitous, and Trustworthy Computing (SUTC), 2010 IEEE International Conference on*, pages 19–26, june 2010.

[6] Che-Jung Hsu, Huey-Ing Liu, and Winston K.G. Seah. Opportunistic routing – A review and the challenges ahead. *Computer Networks*, 55(15):3592–3603, 2011.

[7] Mikkel Koefoed Jakobsen, Jan Madsen, and Michael R. Hansen. Dehar: A distributed energy harvesting aware routing algorithm for ad-hoc multi-hop wireless sensor networks. In *2010 IEEE International Symposium on a World of Wireless Mobile and Multimedia Networks (WoWMoM)*, pages 1–9, June 2010.

[8] Aman Kansal, Jason Hsu, Sadaf Zahedi, and Mani B. Srivastava. Power management in energy harvesting sensor networks. *ACM Trans. Embed. Comput. Syst.*, 6(4), Sept. 2007.

[9] Emanuele Lattanzi, Edoardo Regini, Andrea Acquaviva, and Alessandro Bogliolo. Energetic sustainability of routing algorithms for energy-harvesting wireless sensor networks. *Computer Communications*, 30(14–15):2976–2986, 2007. Network Coverage and Routing Schemes for Wireless Sensor Networks.

[10] Longbi Lin, N.B. Shroff, and R. Srikant. Asymptotically optimal energy-aware routing for multihop wireless networks with renewable energy sources. *IEEE/ACM Transactions on Networking*, 15(5):1021–1034, Oct. 2007.

[11] Longbi Lin, Ness B. Shroff, and R. Srikant. Asymptotically optimal energy-aware routing for multihop wireless networks with renewable energy sources. *IEEE/ACM Trans. Netw.*, 15(5):1021–1034, Oct. 2007.

[12] Jun Lu, Shaobo Liu, Qing Wu, and Qinru Qiu. Accurate modeling and prediction of energy availability in energy harvesting real-time embedded systems. In *2010 International Green Computing Conference*, pages 469–476, Aug. 2010.

[13] Donggeon Noh, Junu Kim, Joonho Lee, Dongeun Lee, Hyuntaek Kwon, and Heonshik Shin. Priority-based routing for solar-powered wireless sensor networks. In *2nd International Symposium on Wireless Pervasive Computing, ISWPC'07*, Feb. 2007.

[14] J.R. Piorno, C. Bergonzini, D. Atienza, and T.S. Rosing. Prediction and management in energy harvested wireless sensor nodes. In *1st International Conference on Wireless Communication, Vehicular Technology, Information Theory and Aerospace Electronic Systems Technology, Wireless VITAE 2009*, pages 6–10, May 2009.

[15] Vijay Raghunathan, A. Kansal, J. Hsu, J. Friedman, and Mani Srivastava. Design considerations for solar energy harvesting wireless embedded systems. In *Fourth International Symposium on Information Processing in Sensor Networks (IPSN 2005)*, pages 457–462, April 2005.

[16] Gunnar Schaefer, Franois Ingelrest, and Martin Vetterli. Potentials of opportunistic routing in energy-constrained wireless sensor networks. In Utz Roedig and Cormac Sreenan,

editors, *Wireless Sensor Networks*, volume 5432 of Lecture Notes in Computer Science, pages 118–133. Springer, Berlin/Heidelberg, 2009.

[17] W.K.G. Seah, Zhi Ang Eu, and Hwee-Pink Tan. Wireless sensor networks powered by ambient energy harvesting (WSN-heap) – Survey and challenges. In *1st International Conference on Wireless Communication, Vehicular Technology, Information Theory and Aerospace Electronic Systems Technology (Wireless VITAE2009)*, pages 1–5, May 2009.

[18] Thomas Watteyne. EXWSN: Experimenting with wireless sensor networks using the ez430-rf2500, 2009.

[19] Kai Zeng, Kui Ren, Wenjing Lou, and Patrick Moran. Energy aware efficient geographic routing in lossy wireless sensor networks with environmental energy supply. *Wireless Networks*, 15:39–51, 2009. 10.1007/s11276-007-0022-0.

3

An Off-line Tool for Accurately Estimating the Lifetime of a Wireless Mote

Sankarkumar Thandapani and Aravind Kailas

Department of Electrical and Computer Engineering, University of North Carolina at Charlotte, Charlotte, NC 28223-0001, USA
e-mail: aravindk@ieee.org

Abstract

Wide scale deployment of battery-backed devices would benefit immensely from ball-park estimates of the energy consumption, and hence their operation lives during different typical operations that would preclude unexpected decreased network coverage as a result of untimely mote "deaths". This chapter presents some preliminary results on profiling the energy consumption in ZigBee-compatible motes during routine communication operations for a point-to-point and star topology network. A simple, intuitive analytical model is presented and is validated using our measurements from our testbed.

Keywords: energy consumption models, platform modeling and analysis tool, performance evaluation and analysis of sensor networks.

3.1 Introduction

Profiling the energy consumption in wireless motes is of paramount importance when modeling the operation life of a network [1]. As an example, inaccuracies in estimating energy consumption led to unexpected mote failures shrinking the size of the habitat monitoring network by almost 50% [2]. Hence, evaluating the energy consumption in motes prior to deploy-

Fabrice Theoleyre and Ai-Chun Pang (Eds.), Internet of Things and
M2M Communications, 49–65.

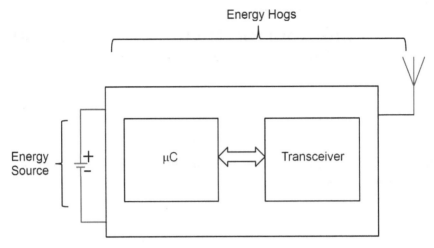

Figure 3.1 High-level wireless mote architecture.

ment is extremely important to avoid the cost associated with unexpected replacement of failed motes. This chapter presents a simple, novel, system-level "off-line" tool that models the energy consumption exclusively in the physical (PHY) layer, (i.e., the transceiver (or the radio) and the μC) along with a cross-layer energy consumption profiling involving the medium access control (MAC) layer during typical mote operations. In areas involving commercial and scientific applications, low-power networks that operate in the industrial/scientific/medical (ISM) bands of 2.4 GHz are being widely adopted. With this in mind, the effectiveness of the proposed analytical energy consumption model has been validated using experimental test beds using commercial ZigBee-ready motes.

3.2 Related Works

Although recent research activities provide many system-level energy consumption models, the accuracy and completeness of the models still remains an open topic of research. For instance, a system-level energy consumption model based on the transceiver battery life has been presented in [3], however, the model did not account for the power consumption in the modulator, filters, analog to digital converter (ADC), or the digital to analog converter (DAC). Another recent energy model took into consideration most of the radio frequency front-end (RF FE) blocks with the exception of the pulse shaping filter because it is usually very low relative to the other "energy-hogs" [4]. We refer to this model as Li's model. However, the model also did not account for the energy costs associated with other baseband blocks such as modulators and coders. Researchers in [5] developed an energy model for low-power wireless motes to analyze the best modulation technique and transmission strategy to minimize energy consumption. We refer to this model as Cui's model. However, the energy costs associated with the modulation techniques were not considered; in this chapter, it is shown that when the modulation technique was considered, the architecture of the transceiver changed and increased the energy consumption by approximately 100%. Furthermore, the power consumption of the μC was not considered in [5] as it was assumed to be negligible. In contrast, our energy consumption model considers the energy consumption in the μC, which in turn increased the accuracy of predicting the operation life of the mote. TOSSIM (TinyOS SIMulator), a simulator for TinyOS networks is used to profile the energy consumption in μC by eliminating inaccurate estimation of assembly language instructions [6].

It has been shown that the energy consumption due to the MAC depends on the number of motes and the beacon interval. However, the energy consumption models involving the MAC layer have not been investigated in the above mentioned works. By analyzing the energy consumption in the key modes of a typical MAC (i.e., Sleep, Direct, etc.), the proposed model accounts for the "energy" costs as a result of the MAC layer. To summarize, the energy profiling of the μC is important to accurately estimate the energy consumption in PHY layer. The model proposed takes into consideration the effects of the modulation technique along with the baseband processing blocks, the μC, and a simple MAC layer resulting in a more accurate and complete estimation of the operation life of a mote during typical operations.

3.3 The Energy Consumption Model

MICAz motes have been used to validate the proposed model because of their wide popularity and extensive use in the sensor network community [7]. MICAz motes operate on 2.4 GHz, and support data rates of up to 250 kbps [8], [9]. The operating voltage range is $2.1 - 3.6$ V [9], and traditionally has four distinct operation modes depending upon the power requirements. A high level system architecture is shown in Figure 3.2. But, for an accurate analysis of the PHY layer energy model a system-on-chip architecture needs to be considered. The proposed energy consumption model has been adapted to the on-chip radio architecture of CC2420 as illustrated in Figure 3.3, the transceiver on MICAz [8], [9] to accommodate the preferred modulation technique, offset-quadrature phase shift keying (O-QPSK). Furthermore, the energy consumption models for the standard blocks (from the state-of-the-art) have been tweaked to incorporate them into the on-chip radio architecture. Parameters such as voltage, device dimensions, frequency, data rate are common to all the components in chip. Since these parameters do not vary for each block, they can be considered as constant and the expression becomes a function of the variables.

As an example, the mathematical equation for the power consumption in a power amplifier (PA) is given by $P_{PA} = \left[\left(\frac{\xi}{\eta} \right) - 1 \right] P_{out}$, where ξ is the peak to average ratio, η is the efficiency, and P_{out} is the output power. Since ξ and η are constant, for a PA, they can be lumped into a PA-specific parameter, denoted by α. Mathematically, $\alpha = \left[\left(\frac{\xi}{\eta} \right) - 1 \right]$. Now, the above expression can be re-written as $P_{PA} = \alpha P_{out}$, where α is a constant that depends on efficiency of amplifier and peak to average ratio and P_{out}. As another example, the conventional equation to calculate the power consumed by a filter is given by $P_{Filter} = SNR^2 * K * frequency * T * Q$, where SNR is the signal to noise power ratio, K is Boltzmann's constant, T is the temperature and Q is the quality factor. In the above equation, since K, T, frequency, and Q are constant, for a filter, they can be lumped into a filter-specific parameter, denoted by β. Mathematically $\beta = K * frequency * T * Q$. Now, the above expression can be re-written as $P_{Filter} = \beta SNR^2$, where β is a constant that depends on frequency, Boltzmann's constant, temperature, and the quality factor of the filter. In other words, the mote-specific components of the energy models for the on-board elements have been lumped into a parameter that is independent of the frequency of operation, radiated power, and data rate.

In this chapter, the energy consumption in the PHY and MAC layers of the mote are modeled for typical operations. The values estimated using the proposed model for power consumption during transmission and reception modes are 70 mW and 67 mW, respectively.

3.3.1 PHY Layer of Wireless Mote

The system-level (as illustrated in Figure 3.2) breakdown of the energy consumption in the radio frequency (RF) transceiver is given by

$$E_{Transceiver} = \sum_i E_{FE,i} + E_{BE,i}, \tag{3.1}$$

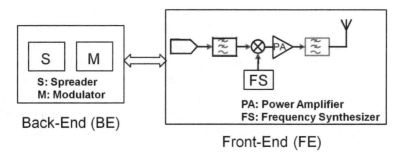

Figure 3.2 Simplified energy model.

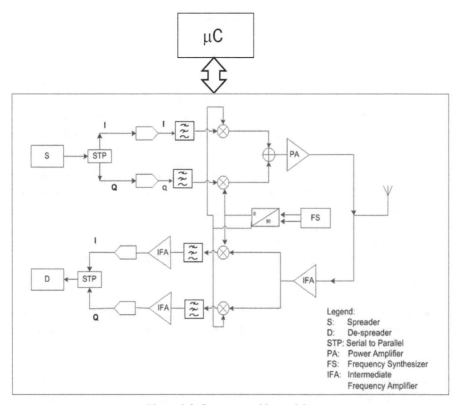

Figure 3.3 System-on-chip model.

where $i \in \{\text{Transmitter}, \text{Receiver}\}$, and E_{FE} and E_{BE} denote the energy consumptions in the front- and back-ends, respectively. The RF FE of a transmitter comprises a DAC, low-pass filter (LPF), mixer, power amplifier (PA), and the back-end (BE) is mainly the digital spreader. Similarly, the RF FE of the receiver consists of an analog to digital converter (ADC), intermediate frequency amplifier (IFA), band pass filter (BPF), mixer, and a digital despreader as the BE block. The frequency synthesizer (FS) is common to both the transmitter and receiver architectures. Therefore, a simple unified expression for the energy consumption in the transceiver, is given by:

$$E_{\text{Transceiver}} = t_{\text{Tx}} \left[\underbrace{P_{\text{Spreader}}}_{\text{BE, Transmitter}} + \underbrace{2P_{\text{DAC}} + 2P_{\text{LPF}} + 2P_{\text{Mixer}} + P_{\text{FS}} + P_{\text{PA}}}_{\text{FE, Transmitter}} \right]$$

$$+ t_{\text{Rx}} \left[\underbrace{P_{\text{Despreader}}}_{\text{BE, Receiver}} + \underbrace{2P_{\text{ADC}} + 2P_{\text{IFA}} + 2P_{\text{Mixer}} + P_{\text{FS}} + P_{\text{LNA}}}_{\text{FE, Receiver}} \right],$$

$$(3.2)$$

where P_{LPF} is the power consumed by the low pass filter, P_{FS} is the power consumed by the frequency synthesizer, t_{Tx} and t_{Rx} are the time durations during which the mote is operating in the transmitting or the receiving mode, respectively. Next, the simplified analytical models for the principle "power hogs" are listed:

- *Power amplifier*: $P_{PA} = \alpha P_{\text{out}}$, where α is a constant that depends on efficiency of amplifier and peak to average ratio and P_{out} is the output power [5].

- *RF Filter*: $P_{\text{Filter}} = \beta \text{SNR}^2 \text{BW}$, where SNR is the signal to noise power ratio, β is a constant that depends on Boltzmann's constant, the temperature, and BW is the bandwidth of operation [4].

- *Low noise amplifier*: $P_{\text{LNA}} = \gamma \frac{A}{\text{NF}}$, where γ is the proportionality constant, A is the gain of the low noise amplifier, and NF is the noise figure [4].

- *Intermediate frequency amplifier*: $P_{\text{IFA}} = \delta(\text{BW} + f_0)\sqrt{\alpha_{\text{BA}}}$, where δ is a coefficient which depends on the device dimensions and process parameters, BW is the bandwidth of the baseband amplifier, f_0 is the center frequency, and α_{BA} is the baseband amplifier gain [4].

- *Spreader (and Despreader)*: $P_{Spreader} = P_{XOR} + NP_{SR}$, where P_{XOR} is the power consumption of the XOR gate, P_{SR} is the power consumption of the shift register, and N is the number of shift registers [10].

3.3.1.1 The μC

The μC on a MICAz mote is ATmega128L, a low-power 8-bit μC [11]. The energy consumption of a μC can be modeled by the following equation:

$$E_{\mu C} = \frac{I * V * N}{f}, \tag{3.3}$$

where I is the current supply, V is the voltage supply, N is the number of cycles, and f is the frequency of operation. The operating system on the MICAz is TinyOS. TinyOS is a free and open source component-based operating system and platform targeting wireless sensor networks. TinyOS applications are written in nesC, an extension to the C programming language designed to embody the structuring concepts and execution model of TinyOS. The nesC code has been converted into assembly language code, a low-level programming language for μCs using XATDB (the graphical symbolic debugger for sensor network simulator), in order to find the energy consumed by the μC while running the executable code [12]. Figure 3.4 illustrates an example computation of the clock cycle count. Consider a+b, a nesC code that performs addition operation between variables a and b, and stores the updated value as a. This is converted into assembly language code ADD Rd,Rr. Two general purpose registers (Rd and Rr) are used in the computations. This code adds the contents in register Rr to the contents in register Rd and stores the sum in register Rd. The ADD command requires 1 clock cycle to execute and this information is available in the datasheet. A Similar procedure is used to compute the clock cycle count of each piece of code and the total number of cycle counts is calculated by summing each instruction's clock cycle. We have calculated the cycle count using XATDB and substituted in equation (3.3). Data such as the supply voltage, supply current, and frequency of operation have been taken from the ATmega 128L data sheet [16].

3.4 MAC Layer of a Wireless Mote

The transceiver for the MICAz is CC2420 that executes the ZigBee stack and operates on IEEE 802.15.4 standard [11]. The IEEE 802.15.4 standard

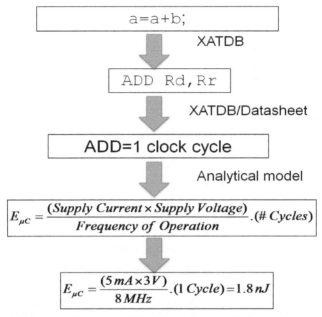

Figure 3.4 Conversion of high level source code into cycle count.

supports two kinds of modes: beacon-enabled and non-beacon-enabled [13]. In a beacon-enabled mode, the motes synchronize with each other and transmit only during their specified beacon [13]. In a non-beacon-enabled mode carrier sense multiple access with collision avoidance (CSMA/CA) is used in order to avoid collision of the packets [13].

3.4.1 CSMA/CA

CSMA/CA is a data collision avoidance protocol for carrier transmission in the IEEE 802.15.4 compatible network. In CSMA/CA, as soon as a mote receives a packet that needs to be transmitted, it checks if the channel is available (i.e., no other mote is transmitting at the time), and transmits it. If the channel is busy, the mote waits for a randomly chosen period of time, and then checks again to see if the channel is available. This period of time is called the back-off factor, and is counted down by a back-off counter. If the channel is clear when the back-off counter reaches zero, the motes transmits the packet. If the channel is not clear when the back-off counter reaches zero, the back-off factor is set again, and the process is repeated.

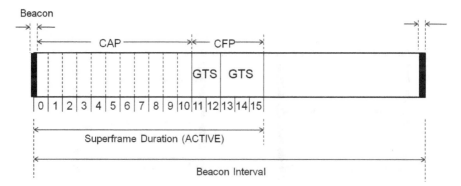

Figure 3.5 Superframe structure in a beacon-enabled mode.

Contention access period (CAP) is the time interval during which the co-ordinators listens to the channel during the whole CAP to detect and receive any data from their daughter motes. The daughter motes may only transmit data and receive an optional acknowledgement (ACK) when needed. In star networks, a device may obtain better quality-of-service (QoS) by the use of guaranteed time slot (GTS), since contention and collisions are avoided. The superframe duration (SD) is the time interval between two superframes. Similarly, the beacon interval (BI) is the time duration between two beacons. Figure 3.5 illustrates the superframe structure in a beacon-enabled mode.

3.4.2 Cross-Layer Energy Profiling

The operation of the network can be broken down into four major modes. Beacon, Direct, Indirect, and Sleep modes. The energy consumption in the MAC layer can thus be modeled as

$$E_i = [P_{i,\text{Beacon}} \cdot t_{i,\text{Beacon}}] + [P_{i,\text{Direct}} \cdot t_{i,\text{Direct}}] + [P_{i,\text{Indirect}} \cdot t_{i,\text{Indirect}}]$$
$$[P_{i,\text{Sleep}} \cdot t_{i,\text{Sleep}}], i \in \{\text{mote, coordinator}\}, \tag{3.4}$$

where P_{Beacon} is the power consumed in beacon mode; t_{Beacon} is the time taken to transmit/receive beacon; P_{Direct} is power consumed in the direct mode; t_{Direct} is time taken to transmit/receive packets during the respective time slot; P_{Indirect} is power consumed in the indirect mode; t_{Indirect} is time taken to transmit/receive packets if the respective time slot is missed; P_{Sleep} is power consumed in the sleep mode; and t_{Sleep} is the mote's sleep time.

In the Beacon-enabled mode the transmitted beacon by the coordinator is received by the mote during this mode. In the Direct mode as shown in

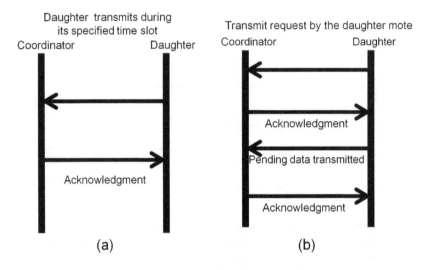

Figure 3.6 (a) Direct mode data, and (b) indirect mode data communication.

Figure 3.6(a), the mote exchanges data with the coordinator in its specified beacon slot. In the Indirect Mode as illustrated in Figure 3.6(b), downlink data from a coordinator to its daughter mote are sent indirectly requiring totally four transmissions. The availability of pending data is signalled in beacons.

3.5 Results and Discussions

3.5.1 Experimental Test-Bed for Analyzing the Energy Consumption in the PHY Layer

The energy consumption due to the PHY layer of the mote has been experimentally verified. Figure 3.7(a) shows the experimental setup. A payload of
20 bytes was transmitted by the source every 250 ms, and the base station received the payload at intervals of 250 ms. Figure 3.7(b) illustrates the timing
diagram that shows the spike during transmission. A sampling rate of 100 Hz
was chosen to log the voltage and current consumption at the two motes to
calculate the power consumption. The data was sampled till the mote died and
the power consumption over the whole test period was averaged to calculate
the power consumption in the mote. The wireless motes were powered by
two AA-sized batteries, each rated at 3000 mAh [14], which when multiplied
with the operating voltage yielded the initial residual energy, which in turn

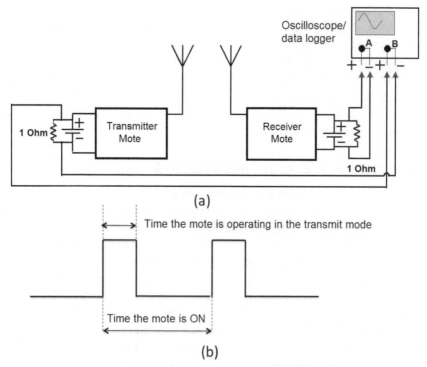

Figure 3.7 (a) Illustration of the experimental setup for analyzing the current and voltage consumptions. (b) Timing diagram for the MICAz motes.

was used in the estimation of the mote operation life. The residual energy of the energy storage device, i.e., the rechargeable battery is computed as:

$$E_{\text{Residual}} = \text{mAh rating} \times \text{voltage across the battery} \tag{3.5}$$

3.5.2 Experimental Test-Bed for Profiling the Energy Consumption in the PHY-MAC Layers

In order to evaluate the effects of the MAC layer on energy consumption in the mote, the following experimental test bed was developed. The test bed consisted of five MICAz motes (i.e., daughter motes) connected as a star network topology connected to a "coordinator" MICAz mote. A round robin scheme was devised for each daughter mote to exchange data with the coordinator. Each mote sent a payload to the coordinator every 250 ms. The coordinator received data from all five daughter motes, and operated at a duty cycle much more than that a daughter mote. The test was executed over seven days during which the voltages and the currents for the daughter and coordinator motes were measured every two hours using voltmeter and ammeter, respectively. The power consumption was calculated by averaging the data collected over the duration of the experiment which in our case was seven days. The measured power consumption value varied and the estimated value was constant. The measured power consumption was averaged in order to obtain a normalized value that can be compared with the estimated value.

3.5.3 PHY Layer Energy Consumption Analysis

MATLAB was used to implement the analytical model. The parameters for the transceiver model were selected from the CC2420 data sheet. The mote operation lives were compared to the experimentally measured values to validate the analytical model. From Figures 3.8 (a) and (b) it can be observed that the proposed model is more accurate than the other state of the art models. The higher accuracy can be explained by the inclusion of the effects of digital modulation technique in the energy consumption computations along with the energy costs associated with the spreader (and despreader) and μC. From Figures 3.8 (a) and (b), it can also be seen that the estimated operation-lives are between 14 and 20% of the experimentally obtained values during the transmission and reception modes, respectively. These differences can be explained by the presence of light-emitting diodes (LEDs), battery leakage, on-board passive elements (i.e., resistors and capacitors) and voltage regulator. The linear curve-fit for the energy consumption in the transmitter is given by:

$$\mathcal{R} = -33.678t + c, \tag{3.6}$$

where $\mathcal{R}$ is the current residual energy on the battery, -33.678 is the slope of the curve plotted, t is the operation hours , and c is 8996.9, the total residual energy on the battery. The slope is the parameter that produces the change in the curve since the residual energy is constant for other models. Similarly, the model for the energy consumption in the receiver is given by:

$$\mathcal{R} = -26.332t + c. \tag{3.7}$$

The slopes in equations (3.6) and (3.7) are functions of the power consumed in the respective modes (transmit or receive mode). It can be inferred from the negative slope that the residual energy on the battery decreases linearly, and can be seen in Figures 3.8 (a) and (b). The *perfectly* linear trends in

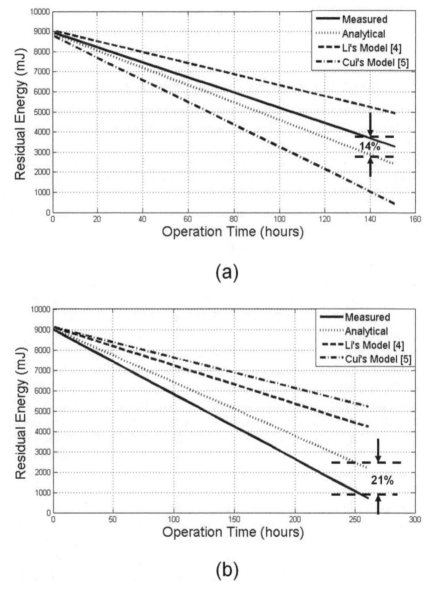

Figure 3.8 (a) Transmitter residual energy versus mote operation life (in hours), and (b) receiver residual energy versus mote operation life (in hours).

Figures 3.8 (a) and (b) are a result of excluding the battery self-discharge from the energy consumption model. The battery self-discharge can be modeled as an exponentially decaying function, but is negligible (typically, 2–3% per month) compared to the reduction in the residual energy as a result of routing the payload.

3.5.4 Cross-Layer Energy Consumption Analysis

In order to operate in the Beacon-enabled mode, the daughter mote receives the beacon and exchanges the data with the coordinator. Based on our analysis of the Beacon-enabled mode, the energy consumption in a daughter mote is dominated by the Direct and Indirect modes. Of the two, the Indirect mode consumes more energy ($\sim$10% more) because the daughter mote performs two additional operations (e.g., sending transmit request, receiving acknow-ledgement, etc.) to exchange data with the coordinator. A daughter mote operating in an Indirect mode is dependant on the packet collision probability, and this has been accounted in the proposed model. A daughter mote may miss a beacon frame resulting in transmitting requests to the coordinator due to unexpected packet collision, and hence, the packet collision probability is taken into account. The MAC layer energy consumption model has been sim-ulated using MATLAB. Equation (3.4) is the analytical model implemented in our MATLAB simulation. Figure 3.9 lends more insights to comparison between measured and simulated values for the daughter and the coordinator motes. The residual energy on the motes are plotted in Figure 3.9 as a function of the operation hours, and is generated in a manner similar to the PHY layer plots. The energy consumption of each coordinator mote is 1.25% of the daughter mote, and this should not be surprising because they operate at a higher duty cycle. From Figure 3.9, it can be observed that the predicted operation lives of the daughter and the coordinator motes obtained using the analytical model were between 15 and 21% of the experimentally measured values, respectively. The errors can be attributed to the exclusion of the en-ergy consumption during transient switching, i.e., Idle to Transmit, Sleep to Idle, etc. These errors can also be attributed to the energy costs associated with the on-board components. Furthermore, the network scanning operation has not been incorporated into our analysis. The daughter mote scans the network to associate with a coordinator, and energy is consumed during this operation has not been considered. However, because the energy consumption during Direct and Indirect modes are more than during the network scanning operation, our assumptions are justified.

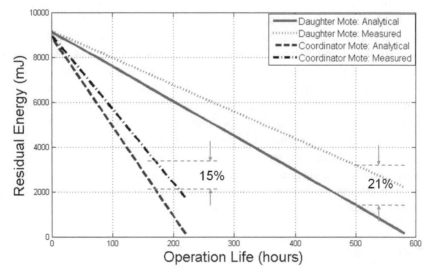

Figure 3.9 Operation lives of the daughter and coordinator motes.

3.6 Conclusions

Wireless motes are promising candidates to enable a plethora of futuristic applications in the commercial world (e.g., structural health monitoring, smart homes, body area networks, to name a few). The wide scale deployment of wireless motes hinges on accurately estimating their operation lives, and this entails modeling the energy consumption accurately, especially for routine operations (such as transmission, reception, sleep, etc.). Erroneous models and estimations are non-reversible and highly expensive, thereby motivating further research in this area. The mote operation lives predicted by the proposed system-level energy consumption model were found to be within 14–21% of the measured values. The higher accuracy relative to the state-of-the-art stems from the inclusion of the energy costs associated with the on-board functions such as μC and digital baseband processing (such as modulation, demodulation, spreading, and despreading). Furthermore, the impact of a simple, practical MAC protocol on the PHY layer provided insights into the energy consumed in Direct, Indirect and Beacon-enabled mode while executing the MAC functionalities. Based on the proposed model, it was seen that the Indirect mode consumed the most energy owing to packet collisions. The probability of packet collisions will scale with the number of motes in the network implying an increase in the total energy consumption of the network. The energy consumption in the coordinator and daughter motes differs due to the increased duty cycle of the coordinator. For future work, it is planned to include an embedded systems/computer architecture analysis (i.e., the `Read` and `Write` functions on the memory needs to be analyzed from the processor's energy profile point of view since the `Read` and `Write` functions consume non-negligible energy and including them will provide a more accurate model). Yet another extension to this work could include adding the effects of instruction pipelining different from the current assumption of sequential execution of instructions in the processor. It would be also be interesting to investigate a dynamic network topology by incorporating mobile motes. Our current analysis of MAC is based on a single hop network, and a multi-hop network is yet to be studied to gain insights into the effects of more complicated routing protocols.

References

[1] O. Landsiedel, K. Wehrle, and S. Gotz. Accurate prediction of power consumption in sensor networks. In *Proc. Second IEEE Workshop Embedded Networked Sensors (EmNetS-II)*, pp. 37–44, May 2005.

[2] R. Szewczyk, J. Polastre, A. Mainwaring, and D. Culler. Lessons from a sensor network expedition, pp. 307–322, 2004.

[3] A. Wang and C. Sodini. A simple energy model for wireless microsensor transceivers. In *Proc. IEEE Global Telecommun. Conf. (GLOBECOM)*, Vol. 5, pp. 3205–3209, Nov.–Dec. 2004.

[4] Y. Li, B. Bakkaloglu, and C. Chakrabarti. A system level energy model and energy-quality evaluation for integrated transceiver front-ends. *IEEE Trans. Very Large Scale Integr. (VLSI) Syst.*, 15(1):90–103, Jan. 2007.

[5] S. Cui, A. Goldsmith, and A. Bahai. Energy-efficiency of mimo and cooperative mimo techniques in sensor networks. *IEEE J. Sel. Areas Commun.*, 22(6):1089–1098, Aug. 2004.

[6] V. Shnayder, M. Hempstead, B. Chen, G. Allen, and M. Welsh. simulating the power consumption of large-scale sensor network applications. In *Proc. 2nd Int. Conf. on Embedded Netw. Sensor Syst.*, pp. 188–200, 2004.

[7] E. Capo-Chichi, H. Guyennet, J. Friedt, I. Johnson, and C. Duffy. Design and implementation of a generic hybrid wireless sensor network platform. In *Proc. 33rd IEEE Conf. on Local Comput. Netw.*, pp. 836–840, Oct. 2008.

[8] N.-J. Oh and S.-G. Lee. Building a 2.4-GHz radio transceiver using IEEE 802.15.4. *IEEE Circuits and Devices Mag.*, 21(6):43–51, Jan.–Feb. 2005.

[9] CC2420 Datasheet. [Online] Available: `http://www.ti.com/lit/ds/symlink/cc2420.pdf`, 2004 (accessed 10 December 2012).

[10] S. Thandapani and A. Kailas. An accurate energy consumption model for the physical layer in a wireless mote. In *Proc. 1st Int. Workshop on Novel approaches to Energy Measurement and Evaluation in Wireless Netw.*, pp. 7850–7854, June 2012.

[11] MICAz Datasheet. [Online] Available PDF: `http://www-db.ics.uci.edu/pages/research/quasar/MPR-MIB\%20Series\%20User\%20Manual\%207430-0021-06A.pdf`, 2004 (accessed 10 December 2012).

[12] J. Polley, D. Blazakis, J. McGee, and D Rusk. ATEMU: A fine-grained sensor network simulator. In *Proc. 1st Annual IEEE Com. Soc. Conf. on Sensor and Ad Hoc Commun.*, pp. 145–152, Oct. 2004.

[13] M. Hännikäinen and T. Hämäläinen. Performance analysis of IEEE 802.15.4 and ZigBee for large-scale wireless sensor network applications. In *Proc. 3rd ACM Int. Workshop on Performance Evaluation of Wireless Ad Hoc, Sensor and Ubiquitous Netw.*, pp. 48-57, 2006.

[14] Energizer Holdings. *Energizer AA Battery Data sheet.* [Online] Available PDF: `http://data.energizer.com/PDFs/191.pdf`, 2010 (accessed 10 December 2012).

[15] G. Bertoni, L. Breveglieri, and M. Venturi. Power aware design of an elliptic curve coprocessor for 8 bit platforms. In *Proc. 4th Annu. IEEE Int. Conf. Pervasive Comput. Commun. Workshops (PerComWorkshops)*, pp. 341–346, Mar. 2006.

[16] ATmega 128L Data sheet. [Online] Available: `http://www.atmel.com/dyn/resources/proddocuments/doc2467.pdf`, 2011 (accessed 10 December 2012).

[17] TinyOS Application Code. [Online] Available: `http://code.google.com/p/tinyos-main/source/browse/\#svn\%2Ftrunk.\%2Fapps\%2FRadioSenseToLeds`, 2010 (accessed 10 December 2012).

Part II

Transmission Scheduling

4

Delay-Constrained Scheduling in Wireless Sensor Networks

Ngoc-Thai Pham, Hoang-Hiep Nguyen, Thong Huynh and
Won-Joo Hwang

*Department of Information and Communications Engineering,
Inje University, Gimhae, Korea
e-mail: yutpham@gmail.com, hoanghiep.hut@gmail.com,
huythongtc@gmail.com, ichwang@inje.ac.kr*

Abstract

In this chapter, we consider the delay guaranteed scheduling and flow control in wireless sensor networks. Delay guarantees are important for delay-sensitive applications in wireless sensor networks, where data delivery is required in a timely manner. For example, packets are dropped if they do not meet a specified deadline. We define an extension of the stability region, namely the constrained stability region, in which the scheduling policy stabilizes the network regarding the delay constraints. Using the stochastic optimization optimal control policies are designed to guarantee delay for each flow in wireless sensor networks. When the arrival rates are within the constrained stability region, the resulting policy is a scheduling policy with delay guarantees. Contrarily, in cases where the arrival rates are outside the constrained stability region, we present a cross-layer design that involves both flow control and scheduling. The resulting policy is a flow control and scheduling policy that guarantees delay constraints and achieves utility performance within $O(1/V)$ of the optimality.

*Fabrice Theoleyre and Ai-Chun Pang (Eds.), Internet of Things and
M2M Communications,* 69–91.

Keywords: scheduling, delay constraint, wireless sensor network, optimal control.

4.1 Introduction

In recent years, Wireless Sensor Networks (WSNs), which consist of sensor nodes which are deployed in the area to monitor, sense and transmit data to the base station have been used for many delay-sensitive application, e.g., video transmission, plant automation and control emergency response, and health care. These applications have many critical QoS requirements, among which meeting delay constraints is an important one. Many WSNs applications require a delay guarantee for time sensitive data. Packets have a specific delay bound, and they are not useful if they are delivered too late. For example, sensor and actor networks require sensors to collect and propagate information in a timely manner so that actors can take timely actions [7, 15]. Another example is water flow monitoring in pipelines in which information of alarming water leakage has to be delivered in a timely fashion by sensor nodes [19]. A target tracking system [20] may require sensors to collect and deliver target information to sink nodes before the target leaves the surveillance field.

Designing a policy for delay-sensitive traffic problems that provide the delay guarantees is desirable in delay-sensitive applications in WSNs [21,25]. It is even more necessary in order to meet the QoS requirement of these services, especially when all sensor devices have limited resource availability such as energy consumption, bandwidth and transmission range, etc. However, the delay is difficult to bound in WSNs due to their unpredictable traffic pattern and the time-varying channel. When the network is synchronized, capturing the delay corresponding to a simple task is done as follows. The sender places a timestamp when sending a packet, the receiver is able to extract the end-to-end delay by also timestamping the packet when it arrives. The difference of the two timestamps is one instance of the delay. However, the use of synchronized clocks is limited in large-scale WSNs due to the large number of message exchanges, and consumes precious energy of the sensor networks. An analytical description of the delay involves parameters from multiple dimensions such as type of network (single hop, multi-hop network), type of target (hop-by-hop delay, end-to-end delay), and type of constraints (mean delay, worst-case delay and low delay). Therefore, a mathematical model is needed in order to capture the performance of the delay in WSNs.

Mathematical models for delay guarantees have been studied in several directions. In the static optimization approach, e.g., under the assumption that the link delay is a static function of the traffic load, the delay constraint on one link is obtained by specifying constraint on the traffic load. However, this assumption is substantially weak in the stochastic environment of wireless networks. In the stochastic network optimization approach [12], the authors studied the stochastic networks with a specific static bound on the queue length. However, it is not enough for a constant bound to capture the dynamic features of the queue length. In another approach, it was shown that the delay in wireless networks can be captured using Little's law [4]. For example, Little's law is used to improve the delay characteristics of TCP traffic [22] or communication over fading channels [3].

In this chapter, we use stochastic network optimization (SNO) framework [6] to analyse the delay as well as the dynamics of wireless environment such as mobility and stochastic link quality. The SNO framework has been recently developed to understand the layering architecture and also to handle the dynamics in wireless environments [5,6]. This framework usually decomposes network architecture into flow control, routing, and scheduling policies. In particular, the authors in [6] developed a set of theories to design and analyse stochastic control policies. These theories include stability region analysis, policy design, and theoretical performance analysis. The scheduling policy follows the Maximal Weighted Matching (MWM) policy [23]. In this seminal work, the authors demonstrated that MWM obtains the largest throughput region. Asymptotic optimality of the throughput and fairness in SNO are guaranteed. However, since queue lengths in the MWM are uncontrollable, the algorithms, which are derived from stochastic network optimization [6], are unable to handle traffic delay so as to satisfy specific QoS requirements.

In this chapter, the delay constraint is transformed to queue length constraint by using Little's law. These queue length constraints are guaranteed using virtual queue length and Lyapunov technique [6]. We design a scheduling policy that guarantees the delay constraints when arrival rates are inside the constrained stability region (Section 4.2). For the case when arrival rates are outside the constrained stability region, we formulate a stochastic optimal control model to obtain utility optimality while guaranteeing delay constraints (Section 4.3). We also show that the joint flow control and scheduling policy achieves optimality within $O(1/V)$ for any control parameter $V > 0$.

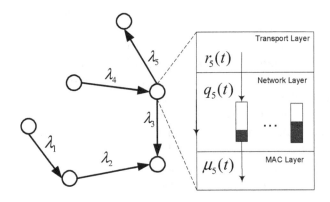

Figure 4.1 System model.

4.2 System Model and Definitions

4.2.1 System Model

We consider a single-hop wireless sensor network with M links, we assume that all sensor nodes receive packets from others after possible retransmissions. Time is divided into equal units $t = \{0, 1, 2 \ldots\}$ called time slots, in e.g., wirelesshart [16] or 802.15.4e [1]. At the beginning of each time slot, the network decides the internal operations including flow control and link scheduling. The proposed system that is shown in Figure 4.1 is modeled by three layers: the transport layer, the network layer, and the MAC layer.

At the MAC layer, let $\mu_m(t)$ be the packet rate on link m at time slot t where the packet rate is total number of data packets per time slot. This means $\mu_m(t) = 1$ if link m is scheduled, otherwise $\mu_m(t) = 0$. For simplicity, we assume that the link qualities independently vary at each time slot according an ON/OFF random process due to the inherent time-varying and mobility of the wireless environment. At each time slot, the selected link is activated to transmit the packet. If it is successful, the sensor node removes the packet from the queue. If it fails, the packet is kept in the queue and will be transmitted in the next time slots. Note that the packet is not dropped when transmission fails. The interference between links imposes a feasible region with an activation vector $\mu(t) = \{\mu_m\}_{m \in \{1, \ldots, M\}}$. Let the feasible region of $\mu(t)$ be Ω, then $\mu(t) \in \Omega$.

At the network layer, let $r(t) = \{r_m(t)\}_{m\in\{1,...,M\}}$ represent the exogenous traffic arriving at time slot t with a mean arrival rate $\lambda_m = E\{r_m(t)\}$ and $r_m(t) \le r_{max}$. Assume that the arrival processes are i.i.d over time and the second moment $E\{r_m^2(t)\}$ of the arrival process is finite. The network layer uses a queue separation scheme where every flow has a separate queue at each node. Let $q_m(t)$ be the queue of flow m, then the dynamic evolution of the queue length is

$$q_m(t+1) = q_m(t) - \mu_m(t) + r_m(t). \tag{4.1}$$

4.2.2 Definitions

In this section, we define basic definitions related to the stability region and the review of existing results.

Definition 1 (Network Stability): A queue q is stable if

$$\limsup_{T->\infty} \frac{1}{T} \sum_{t=0}^{T-1} E\{q(t)\} < \infty.$$

A network is stable if all queues are stable.

Definition 2 (Stability Region): The stability region of a scheduling policy is the set of arrival rates $\{\lambda_m\}_{m\in\{1,...,M\}}$ that stabilize the system under the policy. The union of stability regions of all scheduling policies is the stability region of the system.

Definition 3 (Capacity Region): The capacity region, denoted by Λ, is a set of all arrival rate vectors that can be stably supported by the network.

Definition 4 (Constrained Stability Region): The constrained stability region of a scheduling policy is a set of all arrival rates that stabilizes the system under the scheduling policy. Moreover, the mean delay under the policy satisfies the specified delay constraints. In this case, arrival rates are admissible.

It is clear that the stability region of the system is a subset of the capacity region. In the presence of delay constraints, the *constrained stability region* is a subset of the stability region of the policy and the stability region of system (Figure 4.2). Moreover, as shown in [6,8], since queue length depends

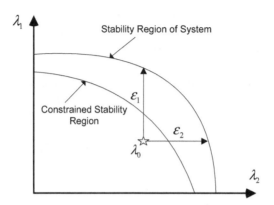

Figure 4.2 Stability region of the system and the constrained stability region.

on the inverse of the distance between the arrival rate and the boundary of stability region, the stricter the delay constraints are, the smaller the constrained stability region is. The queue length primarily depends on arrival and departure traffic distribution. A general stability region must not depend on any assumption of traffic distribution since the assumption on a specific traffic pattern distribution may not hold in all cases. For that reason, queue length bounds should be achieved using the Lyapunov technique [6,8] which is independent from the traffic distribution.

The principle of the Lyapunov technique is briefly described as follows: first define the Lyapunov function as $L(t) = \sum_{m=1}^{M} (q_m(t))^2$, which is the sum square of all queue lengths at time slot t. If $L(t)$ is small, then all queue lengths are small, e.g., if there is a finite constant C such that $L(t) \leq C$ for $\forall t$, then it clearly guarantees that $q_m(t) \leq \sqrt{C}$ for $\forall m$, thus all queue lengths are bounded by $\sqrt{C}$. The condition $L(t) \leq C$ is gradually obtained by minimizing the *Lyapunov drift*, which is defined as $\Delta L(t) = L(t+1) - L(t)$, from one slot to the next, in other words, minimizing the Lyapunov drift will lower the Lyapunov function slot by slot. For more detail, the readers can refer to the Appendix and [6]. The following lemma formulates the constrained stability region in general cases using the Lyapunov technique.

Lemma 1 (*Constrained Stability Region*): $\overrightarrow{\lambda}$ *belongs to the constrained stability region* Λ_c *of the system with delay constraints* $E\{q_m(t)\} \leq \overline{q}_m$ *with* $m \in \{1, \ldots, M\}$ *if there is a vector* $\varepsilon = \{\varepsilon_m\}_{m \in \{1,\ldots,M\}} > 0$ *such that*

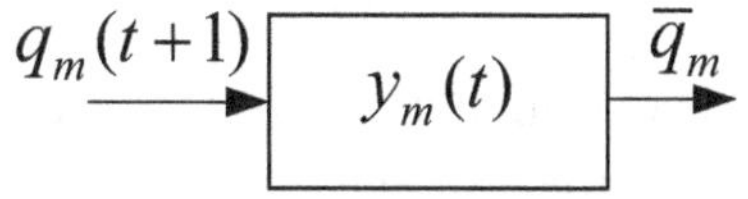

Figure 4.3 Virtual queue with arrival rate $q_m(t+1)$ and service rate $\bar{q}_m$.

$\overrightarrow{\lambda} + \overrightarrow{\varepsilon} \in \Lambda$ *and*

$$\frac{E\{r_m^2(t)\} - 2\lambda_m^2 + \lambda_m}{2\varepsilon_m} \leq \bar{q}_m \; with \; m \in \{1, \ldots, M\}. \tag{4.2}$$

Proof. See Appendix A

4.3 Delay-Constrained Scheduling (DSC)

4.3.1 DSC Problem Formulation

For each flow m, we define a virtual queue $y_m(t)$ with evolution:

$$y_m(t+1) = \max\left(y_m(t) - \bar{q}_m, 0\right) + q_m(t) - \mu_m(t) + r_m(t). \tag{4.3}$$

Since $q_m(t+1) = q_m(t) - \mu_m(t) + r_m(t)$, the virtual queue length $y_m(t)$ can be seen as the queue length with the arrival process $q_m(t+1)$ and the server process $\bar{q}_m$ (Figure 4.3). Thus, if there is a scheduling policy that maintains the stability of the virtual queue, the arrival rate must be smaller or equal to the server rate, i.e., $q_m(t+1) \leq \bar{q}_m$. Hence, the delay constraints hold. We consider a discrete-time linear time-invariant system with dynamics in equations (4.1) and (4.3) where $\Phi(t) = \{q_c(t), y_c(t)\}_{c \in \{1,\ldots,C\}} \geq 0$ are the states, $\mu(t) \in \Omega$ is the control input. Note that only $\mu(t)$ is variable here since only the network and MAC layers are considered in this scheduling policy design. In addition, there is no need to consider the stability of q since the stability of y implies the stability of q. In order to guarantee the delay constraints, we stabilize the virtual queue system by minimizing the following cost function:

$$J = \lim_{T \to \infty} \sup \frac{1}{T} \sum_{m=1}^{M} \sum_{t=0}^{T} E\left\{y_m(t+1)^2\right\}. \tag{4.4}$$

4.3.2 DSC Policy

Here, we have to find the activation vector $\mu(t)$ at each beginning of time slot t to minimize the cost function J, which is the expected sum square of all virtual queues length over horizon T for stabilizing all virtual queues. In general, the linear stochastic optimal control problem can be solved effectively using the optimal policy for only a few special cases [11,24]. However, there exist different methods to find *suboptimal control policy* [24]. Regarding the computational complexity and its compliance to the objective function, we consider the Control-Lyapunov Feedback method, in which the suboptimal control feedback, i.e., the activation vector $\mu(t)$, is chosen such that it minimizes the instantaneous sum square of all virtual queues at each time slot t, i.e.,

$$\mu(t) = \arg\min_{\mu \in \Omega} \left(\sum_{m=1}^{M} y_m(t+1)^2 \right). \tag{4.5}$$

Proposition 1: *The following inequality holds at any time slot t*

$$\sum_{m=1}^{M} y_m(t+1)^2 \leq \sum_{m=1}^{M} y_m(t)^2 + B(t) + D(t)$$

$$+ \left(\sum_{m=1}^{M} 2r_m(t)(q_m(t) + y_m(t)) \right)$$

$$\underbrace{- 2 \sum_{m=1}^{M} (q_m(t) + y_m(t)) \mu_m(t)}_{Scheduling} \tag{4.6}$$

where

$$B(t) = \sum_{m=1}^{M} \bar{q}_m(t)^2 + \sum_{m=1}^{M} (r_m(t) - \mu_m(t))^2$$

and

$$D(t) = \sum_{m=1}^{M} q_m(t)^2 - 2 \sum_{m=1}^{M} y_m(t)(\bar{q}_m - q_m(t)).$$

Proof. Applying Lemma 4.3 from [6] to equation (4.3), summing over all m entries, and re-arranging yields (4.6). □

A suboptimal policy can be derived from equation (4.5) by minimizing the upper bound of the instantaneous sum square of virtual queue length at each time slot t, i.e., minimizing the right-hand side of (4.6). Thus, the delay-constrained scheduling, namely DCS, is derived as follows:

Delay-Constrained Scheduling: *At each time slot, DSC selects an activation vector $\mu(t)$ such that:*

$$\mu(t) = \arg\max_{\mu \in \Omega} \sum_{c=1,...,M} (q_m(t) + y_m(t))\mu_m(t). \tag{4.7}$$

The DSC policy selects the activation vector such that it maximizes the above weighted sum. It has a similar form to the well-known Backpressure algorithm [23] however, the weights of links are calculated differently. In the DSC, the weights take into account the physical queue length and the virtual queue length, i.e., $(q_m(t) + y_m(t))$, compare to only physical queue length, i.e., $q_m(t)$ in Backpressure routing algorithm. Note that the decision of the DSC depends only on the current state of the system, i.e., queue length $q_m(t)$ and the activation vector $\mu_m(t)$.

Theorem 1. *If there is λ such that $\overrightarrow{\lambda} \in \Lambda_c$ then the queueing system is stable and all delay constraints hold under DSC.*

Proof. See Appendix B.

4.3.3 Implementation Issues

- *Admission Control*: Admission control is a necessary component in the delay guarantees implementation to maintain arrival rates inside the constrained stability region. Note that the delay constraints require the virtual queues are bounded. Hence, the admission control may be implemented simply by limiting the upper bound of the virtual queue systems.

- *Distributed Implementation*: Implementation of the DSC scheduling requires global information of the physical queue length, virtual queue length and packet rate vector. However, note that the DSC policy can be implemented in a distributed manner once the packet rate $\mu_m(t)$ on link m at each time slot t is decided. In this case, each sensor node needs only to expose $q'_m(t) = q_m(t) + y_m(t)$ to the external sensor nodes, and the DSC becomes

$$\mu(t) = \arg \max_{\mu \in \Omega} \sum_{m=1,\dots,M} (q'_m(t))\mu_m(t).$$

This scheduling policy has a similar form to that of MWM. The packet rate $\mu_m(t)$ can be determined, e.g., as in [14] based on the signal-to-interference model, where each node randomly sends a pilot signal every slot t, then all potential receiver nodes measure the received pilot signal and feedback the signal-to-interference to the transmitter to decide the packet transmission rate. The complexity of DSC policy is $O(M^3)$. The implementation techniques that are applied to MWM [2], e.g. imperfect scheduling [18], can also be used to implement the DSC.

4.4 Delay-Constrained Flow Control and Scheduling (DFS)

4.4.1 DFS Problem Formulation

When the exogenous arrival rates are outside the capacity region Λ, flow control must be considered to maintain the stability of queues. At the transport layer, assume that there is an infinite-backlogged *transport layer storage*. The flow control admits data packets from the transport layer to the corresponding network queues maintaining the network in the stability region and guaranteeing the delay constraints. Let

$$\bar{r}_m(t) = \frac{1}{t} \sum_{\tau=0}^{t-1} E\{r_m(\tau)\}$$

represent the long term packet rate admitted to queue q_m. This is similar to the arrival rate λ_m, however, we consider $\bar{r}_m$ here as a variable in the control flow. Each packet rate $\bar{r}_m$ is associated with a utility function $U_m(\bar{r}_m)$. The utility functions are continuous, differentiable, strictly concave and increasing. These utility functions are used to obtain the fairness properties as seen in [5, 9].

We consider a discrete-time linear time-invariant system with dynamics in the equations (4.1) and (4.3) where $\Phi(t) = \{q_m(t), y_m(t)\}_{m \in \{1,\dots,M\}} \geq 0$ are states and $\mu(t) \in \Omega$ and $r(t) \in \Lambda$ are the control inputs. $\mu(t)$ and $r(t)$ are variable because both flow control and scheduling policy are included in this design. Since the optimal control problem is to maintain the stability of the dynamic system and to maximize the total utilities $\sum_{m=1}^{M} U_m(r_m)$, we define the cost function of the system as follows:

$$J = \lim_{T \to \infty} \sup \frac{1}{T} \sum_{t=0}^{T} E \left\{ \sum_{m=1}^{M} (y_m(t+1)^2 - V U_m(r_m(t))) \right\}.$$

The first term in function J is the average expectation sum quadratic of the virtual queue length. The second term of J is the average expectation of total utilities. The minimization of J assures that virtual queue length is minimized and total utilities is maximized. V is a control parameter that provides the tradeoff between the stability of queue y and the attained utilities.

4.4.2 DFS Policy

Similar to DSC policy, the suboptimal control policy following the Control-Lyapunov Feedback method is given by

$$\{\mu(t), r(t)\} = \arg \min_{\mu \in \Omega, r \in \Lambda} \left(\sum_{m=1}^{M} y_m(t+1)^2 - V \sum_{m=1}^{M} U_m(r_m(t)) \right). \quad (4.8)$$

Proposition 2: *The following inequality holds at any time slot t:*

$$\sum_{m=1}^{M} y_m(t+1)^2 - V \sum_{m=1}^{M} U_m(r_m(t))$$

$$\leq \sum_{m=1}^{M} y_m(t)^2 + B(t) + D(t)$$

$$\underbrace{- 2 \sum_{m=1}^{M} (q_m(t) + y_m(t)) \mu_m(t)}_{Scheduling}$$

$$\underbrace{- \left(V \sum_{m=1}^{M} U_m(r_m(t)) - \sum_{m=1}^{M} 2r_m(t)(q_m(t) + y_m(t)) \right)}_{Flow\,Control}. \quad (4.9)$$

Proof. Subtracting $V \sum_{m=1}^{M} U_m(r_m(t))$ on both sides of (4.6) yields (4.9). □

The suboptimal control policy (4.8) is derived by minimizing the upper bound of $\sum_{m=1}^{M} y_m(t+1)^2 - V \sum_{m=1}^{M} U_m(r_m(t))$ at each slot, or equivalently, minimizing the left-hand side of equation (4.9).

The derived control policy is decomposed into two tasks: flow control on the transport layer, which controls the arrival rates to the network layer, and the scheduling policy, which determines the activation vector. The DFS policy is defined as follows:

Delay-Constrained Flow Control and Scheduling:

- *Flow control: For every time slot, flow m observes the queue states $\{q_m(t), y_m(t)\}$ and chooses $r_m(t)$ such that it is the solution of:*

$$
\begin{aligned}
&\max_{r_m} \quad V U_m(r_m(t)) - 2r_m(t)(q_m(t) + y_m(t)) \\
&\text{s.t.} \quad 0 \le r_m(t) \le r_{\max}.
\end{aligned}
\tag{4.10}
$$

- *Scheduling: At each time slot t, the policy selects an activation vector $\mu(t)$ such that*

$$
\mu(t) = \max_{\mu \in \Omega} \sum_{m=1,\dots,M} (q_m(t) + y_m(t))\mu_m(t).
$$

The DFS policy has a similar form to that of the SNO algorithm [6] for single-hop networks except that it has an additional virtual queue length for each flow. The flow control solves the optimization problem based only on the local physical queue and the virtual queue information at the source nodes. This is a distributed policy since information involves only the source nodes. The scheduling policy is exactly the DSC policy. The following performance result holds for the DFS policy.

Theorem 2. *If the constrained stability region is non-empty, then for any parameter V, the DFS stabilizes the network, satisfies the delay constraints and yields the following utility bounds:*

$$
\liminf_{t \to \infty} \sum_{m=1}^{M} U_m(\bar{r}_m(t)) \ge \sum_{m=1}^{M} U_m(r_m^*) - \frac{\bar{B} + \sum_{m=1}^{M} \bar{q}_m^2}{V},
$$

where r_m^* is the optimal solution of the delay-constrained optimal control problem and

$$
\bar{r}_m(t) = \frac{1}{t} \sum_{\tau=0}^{t-1} E\{r_m(t)\}.
$$

Proof. See Appendix C.

The DFS attains a solution with a constant gap of $O(1/V)$ to the optimal solution provided that there is an arrival rate that satisfies the delay constraints. By controlling parameter V, we can achieve total utility arbitrarily close to the optimal point.

4.5 Performance Analysis

We study the behavior of the DSC and DFS policies in a star topology network with three links (Figure 4.4(a)). We set up a dynamic wireless environment where the link qualities and the arrival rates are stationary processes. A 2-hop interference model is used to determine feasible activation vectors. At each time slot, our simulation runs the following steps. At the first step, the link quality is generated by a random process with ON probability 0.8. For the simulation for DSC scheduling, the external packet arrival rate at each node is i.i.d with mean λ_m. At the second step, we find activation vector μ according to the DSC scheduling policy and the admitted rate r_m at each node according to the flow control policy based on the interference model, the physical queue q, and the virtual queue y. At the last step, queuing evolution is updated. No packet is discarded during the physical queuing update.

4.5.1 Performance Analysis of the DSC

In the DSC policy, the queue length constraints of node 1, 2, 3 are 2, 4 and 5 respectively. As shown in Figure 4.4(b), although the queue lengths are randomly involved due to the randomness of the link qualities and the arrival rates, the queue length of every link remains near the predefined queue length. In the second scenario, we compare the delay performance of the DSC and the WMW algorithm under the variance of the arrival rates. The arrival rate to each link is changed and the dependency between the delay and the arrival rates is shown (Figure 4.4(c). It can be observed that the results have two interesting features of the DSC. First, when the arrival rates are small, the delay of the DSC and MWM are the same. This is because the queue lengths are small, virtual queue lengths are zero. Thus, both algorithms operate identically. When the arrival rate increases, the DSC acts to reduce either longer queues or the longest queue, thus, the average delay of DSC is smaller than that of MWM. Second, since the DSC arrival packet rate is limited within the

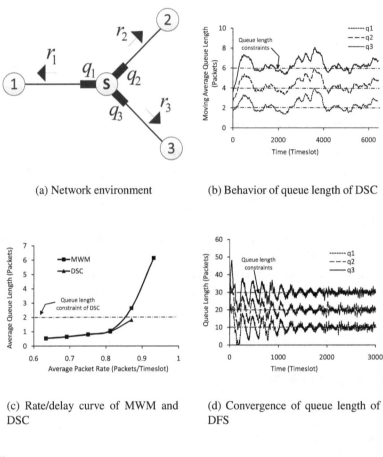

(a) Network environment

(b) Behavior of queue length of DSC

(c) Rate/delay curve of MWM and DSC

(d) Convergence of queue length of DFS

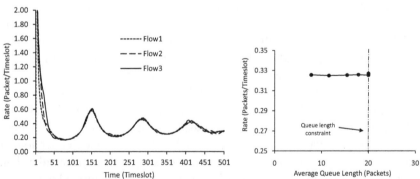

(e) Convergence of packet rate of DFS

(f) Pareto tradeoff curve of DFS

Figure 4.4 Performance analysis of the DSC and DFS.

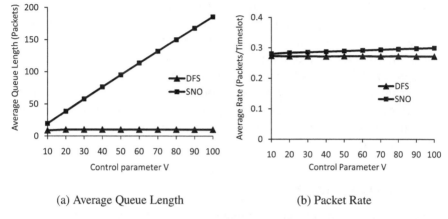

(a) Average Queue Length   (b) Packet Rate

Figure 4.5 Performance comparisons of the DFS and SNO.

constrained stability region to guarantee the delay constraints, by limiting the total virtual queues length below 1000, the rate/delay curve of the DSC stops at around 0.87 (packets/timeslot) while the rate/delay curve of the MWM increases significantly.

4.5.2 Performance Analysis of the DFS

In the DFS policy, the utility function is a $\log(\cdot)$ function. Control parameter V is 100. The queue length constraints on the links are set to 10, 20, and 30, respectively. Figures 4.4(d) and 4.4(e) show the dynamics of the queue length and the admitted rates of the three flows. As predicted in the analytical results, the queue length fluctuates and converges at the predefined queue length while the packet rates of the three flows converge at the same values showing the fairness property of the utility functions. Figure 4.4(f) shows the Pareto tradeoff curve between rates and delay when the control parameter varies from 10 to 100. The curve locates under the queue length average of 20 for all values of the control parameter V. This implies that the delay is guaranteed regardless of the control parameter V.

Finally, we investigate performance of the DFS by performing an exhaustive simulation on a single hop network with grid topology of 24 links and compare the performance with the SNO algorithms. The number of the flows and the position of the flows are randomly generated for each simulation run. For each control parameter value, the simulation is performed 100 times and each time for 30,000 timeslots. The queue length constraints are set to

10 for all links. The simulation results show that the DFS has a persistent delay level below the predefined queue length (Figure 4.5(a)) while the average rate is close to the optimal average rate resulting from SNO algorithms (Figure 4.5(b)). It is important to note that although the queue lengths may be small, the DFS can achieve the average rate that is close to the optimal average rate. This is because the DFS only limits the upper bound of the queue lengths. There are packets ready to transmit in the physical queues under the DFS, therefore the DFS also can attain high average rate under delay constraints.

4.6 Open Challenges

Energy consumption is a very important issue in WSNs as each sensor node has small battery capacity. Some approaches use the topology control and power management algorithms at the MAC layer to increase the life-time of the network [13, 17]. However, this will significantly affect the delay in the network. Thus, the link scheduling and flow control for delay guarantees maybe need to consider the *energy-latency* tradeoffs to prolong the network life expectancy. In such a systems, the problem is more complicated as the latency of packets consists of not only the delay in queues in each sensor node, but also the delay for packet transmission, re-transmissions, channel access and waking up sleeping sensor nodes.

In WSNs, multi-hop transmission is a common mechanism used due to the short-range communication between sensor nodes. In multi-hop WSNs, transmitted packets are relayed through several intermediate nodes to save energy and increase the network life time. However, multi-hop transmission will increase the delay due to queuing and processing at the intermediate nodes. Generally, the larger number of hops, the smaller energy consumption, but also the higher end-to-end delay. Thus, it is necessary to address this tradeoff to ensure the delay constraints in multi-hop WSNs. Current research on delay guarantees for multi-hop wireless networks usually uses the latency delay obtained from single-hop queuing models [10]. However, since this computation only holds under the assumption that the queues in network nodes follow an independent process and link rates follow a specified distribution, it seems unreasonable for general WSNs. On the other hand, the solution for this problem is usually centralized and not adaptive to dynamics of wireless networks.

Another major issue in the WSNs is that the wireless links are usually unreliable. Each transmission link has an associated error probability that may vary with time due to various factors like fading, interference, multipath effects, and collisions. However, the previous work on delay guarantees did not consider the property of the wireless links. On other hand, end-to-end delay is extremely important in path reliability. If we use bad radio links with a large packet error rate, a large number of retransmissions will waste energy and hence, shorten the network lifetime. On the other hand, retransmissions also cause higher collision probability and delivery delay. Thus, we have to model unreliable wireless links to guarantee delay-sensitive traffic.

4.7 Conclusion

This chapter presents a framework for delay-guaranteed scheduling in wireless sensor networks. For delay guarantee scheduling only, the delay constraints hold under the DSC policy provided that arrival rates are within the constrained stability regions. For scheduling and flow control, the theoretical analysis and simulation show that the utility is asymptotically close to the optimality while the delay constraints hold. In this chapter, the storage in each sensor node is assumed infinite, since real-time applications, e.g., video streaming, may have finite-backlogged storage, one can consider using the *auxiliary variable* technique [6] to remove this assumption.

Appendix A – Proof of Lemma 1

Proof. To begin the proof, we first compute the bound of a queue with arrival rate λ_m and departure rate $\mu_m = \lambda_m + \varepsilon_m$. Consider the Lyapunov function $L(y_m(t)) = y_m(t)^2$ of dynamic evolution (4.1). The drift of the Lyapunov function is defined as $\Delta L(y_m(t)) = y_m(t+1)^2 - y_m(t)^2$. Squaring both sides of (4.1) and rearranging, we have

$$\Delta L(y_m) = \bar{B}_m - 2E\{q_m(t)(\mu_m(t) - r_m(t))\} \qquad (4.11)$$

where $\bar{B}_m = E\{(\mu_m(t) - r_m(t))^2\}$.

Using the facts that $E\{\mu_m^2\} = E\{\mu_m\}$ and $E\{\mu_m\} = \lambda_m$ under the stability condition we obtain:

$$\bar{B}_m = E\{r_m(t)^2\} - 2\lambda_m^2 + \lambda_m.$$

Using $\mu_m = \lambda_m + \varepsilon_m$ in (4.9) yields

$$E\{\Delta L(y_m)\} = \bar{B}_m - 2\varepsilon_m E\{q_m(t)\}.$$

Using lemma 4.2 from [6] with the above condition, the following is obtained:

$$E\{q_m(t)\} \leq \frac{E\{r_m^2(t)\} - 2\lambda_m^2 + \lambda_m}{2\varepsilon_m}. \tag{4.12}$$

From (4.2) and (4.12), it can be concluded that Lemma 1 holds.   □

Appendix B – Proof of Theorem 2

Proof of this theorem uses similar techniques to those of theorem 4.5 in [6]. Taking the expectation of (4.3) yields

$$\Delta L(y) \leq \bar{B} + \sum_{m=1}^{M} E\{q_m(t)^2\} - 2\sum_{m=1}^{M} y_m(t) E\{(\bar{q}_m - q_m(t))|\Phi(t)\}$$

$$- 2E\left\{ \sum_{m=1}^{M} (q_m(t) + y_m(t))\mu_M(t) \middle| \Phi(t) \right\}$$

$$- 2E\left\{ \sum_{m=1}^{M} r_m(t)(q_m(t) + y_m(t)) \middle| \Phi(t) \right\} \tag{4.13}$$

where

$$\bar{B} = \sum_{m=1}^{M} \bar{q}_m^2 + E\left\{ \sum_{m=1}^{M} (\mu_m(t) - r_m(t))^2 \right\}.$$

Because $\lambda \in \Lambda_c$ and according to corollary 3.9 in [6], there is a randomized algorithm that allocates $\mu(t)$ such that $\lambda + \varepsilon \in \Lambda$ where ε_m satisfies condition (4.2). Hence under this policy we have $\beta = \{\beta_m\}_{m=\{1,\dots,M\}} > 0$ such that $\bar{q}_m = E\{q_m(t)\} + \beta_m$. Thus, the following is obtained:

$$E\left\{ \sum_{m=1}^{M} q_m(t)^2 \right\} - 2\sum_{m=1}^{M} y_m(t) E\{(\bar{q}_m - q_m(t))|\Phi(t)\}$$

$$\leq \sum_{m=1}^{M} \bar{q}_m^2 - 2\sum_{m=1}^{M} \beta_m y_m(t). \tag{4.14}$$

and

$$E\{\mu_m(t)|\Phi(t)\} = \lambda_m + \varepsilon_m \tag{4.15}$$

Using (4.14) and (4.15) in (4.13) yields

$$\Delta L(y) \leq \bar{B} + \sum_{m=1}^{M} \bar{q}_m^2 - 2 \sum_{m=1}^{M} \beta_m y_m(t)$$
$$- 2 \sum_{m=1}^{M} \varepsilon_m [q_m(t) + y_m(t)].$$

Since $q_m(t) \geq 0$, the following result is obtained for the randomized algorithm:

$$\Delta L(y) \leq \bar{B} + \sum_{m=1}^{M} \bar{q}_m^2 - 2 \sum_{m=1}^{M} (\beta_m + \varepsilon_m) y_m(t). \qquad (4.16)$$

Recall that the DSC minimizes the left-hand side of (4.13), thus any scheduling policy should have a greater left-hand side including the randomized algorithm. Thus, (4.16) also holds for the DCS. Using this drift inequality as the condition for lemma 4.2 from [6], we find the virtual queuing system is stable under the DSC. Hence, all the physical queues are bounded. The theorem is proved.

Appendix C – Proof of Theorem 3

The following result is proved first.

Lemma 2. *Under the DFS, the virtual queue system is stable and thus all delay constraints hold.*

Proof. (Lemma 2) Solving problem (4.10) yields the following solution:

$$r_m(t) = \begin{cases} r_{\max} & r_m^* > r_{\max}, \\ r_m^* & otherwise \end{cases}$$

where

$$r_m^* = (U_m')^{-1} \left(\frac{2}{V} (q_m(t) + y_m(t)) \right).$$

Since U_m is strictly concave and increasing, r_m^* goes to zero when $q_m(t) + y_m(t)$ goes to infinity. Thus, there must be $\lambda = E\{r_m^*\} > 0$ such that $\lambda \in \Lambda_c$ and $q_m(t) + y_m(t)$ are bounded. Thus, the lemma holds. $\qquad \square$

Proof. (Theorem 2) The proof of this theorem uses a similar technique as the one in the proof of theorem 5.1 in [6].

Taking the expectation of (4.9) yields

$$\Delta L(y) - E\left\{V\sum_{m=1}^{M}U_m(r_m(t))\right\}$$

$$\leq \bar{B} + \sum_{m=1}^{M}E\left\{q_m(t)^2\right\} - 2\sum_{m=1}^{M}y_m(t)E\left\{(\bar{q}_m - q_m(t))\middle|\Phi(t)\right\}$$

$$- 2E\left\{\sum_{m=1}^{M}(q_m(t) + y_m(t))\mu_m(t)\middle|\Phi(t)\right\}$$

$$- E\left\{V\sum_{m=1}^{M}U_m(r_m(t)) - 2\sum_{m=1}^{M}r_m(t)(q_m(t) + y_m(t))\middle|\Phi(t)\right\}.$$

$$(4.17)$$

Since the system is stable due to Lemma 2 and flow control and scheduling maximize over all the possible choices of $\{q(t), y(t)\}$, there exists an operation point such that the *DFS* flow control yields

$$VU_m(r_m^*(\varepsilon)) - 2r_m^*(\varepsilon)(q_m(t) + y_m(t)) \leq VU_m(r_m) - 2r_m(q_m(t) + y_m(t))$$
$$(4.18)$$

and the scheduling policy yields

$$E\left\{\sum_{m=1}^{M}(q_m(t) + y_m(t))\,\mu_m(t)\,|\Phi(t)\right\}$$

$$\geq \sum_{m=1}^{M}(q_m(t) + y_m(t))\,(r_m^*(\varepsilon_m) + \varepsilon_m),\qquad(4.19)$$

where $r^*(\varepsilon) = \{r_m^*(\varepsilon)\}$ is the arrival rate such that $\varepsilon > 0$ and $r^*(\varepsilon) + \varepsilon \in \Lambda_c$.

Because $\lambda \in \Lambda_c$ due to Lemma 2, all the delay constraints hold. Hence, under the DFS we have $\beta = \{\beta_m\}_{m=\{1,...,M\}} \geq 0$ such that $\bar{q}_m = E\{q_m(t)\} +$

β_m. Thus, we have

$$E\left\{\sum_{m=1}^{M} q_m(t)^2\right\} - 2\sum_{m=1}^{M} y_m(t)E\left\{(\bar{q}_m - q_m(t))|\Phi(t)\right\}$$

$$\leq \sum_{m=1}^{M} \bar{q}_m^2 - 2\sum_{m=1}^{M} \beta_m y_m(t). \tag{4.20}$$

By using (4.18), (4.19), and (4.20) in (4.17) and by rearranging and omitting $q_m(t) \geq 0$, the following is obtained:

$$\Delta L(y(t)) - V\sum_{m=1}^{M} U_m(r_m(t))$$

$$\leq \bar{B} + \sum_{m=1}^{M} \bar{q}_m^2 - 2\sum_{m=1}^{M} (\beta_m + \varepsilon_m)y_m(t) - VU_m(r_m^*(\varepsilon)).$$

The above drift inequality has the form of the condition specified in [6, theorem 5.4]. Thus, we have

$$\liminf_{t\to\infty} \sum_{m=1}^{M} U_m(\bar{r}_m(t)) \geq \sum_{m=1}^{M} U_m(r_m^*(\varepsilon)) - \frac{(\bar{B} + \sum_{m=1}^{M} \bar{q}_m^2)}{V}.$$

By choosing ε close to the boundary of the constrained stability region, $r^*(\varepsilon)$ can be obtained close to the optimal operation point r^* or equivalently we have the specified utility bound. This proves Theorem 2. $\qquad\square$

References

[1] IEEE standard for local and metropolitan area networks – Part 15.4: Low-rate wireless personal area networks (LR-WPANS) amendment 1: Mac sublayer. *IEEE Std 802.15.4e*, 2012.

[2] D. Avis. A survey of heuristics for the weighted matching problem. *Network*, 13(4):475–493, 1983.

[3] R. A. Berry and R. G. Gallager. Communication over fading channels with delay constraints. *IEEE Transactions on Information Theory*, 48(5):1135–1149, May 2002.

[4] D. Bertsekas and R. Gallager. *Data Network*. Prentice Hall, 1992.

[5] A. Eryilmaz. Fair resource allocation in wireless networks using queue-length-based scheduling and congestion control. In *Proceedings IEEE INFOCOM 2005*, pages 1794–1803, 2005.

[6] L. Georgiadis, M. J. Neely, and L. Tassiulas. *Resource allocation and cross-layer control in wireless networks.* Found. Trends Netw, 2006.

[7] V. C. Gungor, O. B. Akan, and I. F. Akyildiz. A real-time and reliable transport (RT)2 protocol for wireless sensor and actor networks. *IEEE/ACM Transactions on Networking*, 16(2):359–370, April 2008.

[8] K. Kar, Xiang Luo, and S. Sarkar. Delay guarantees for throughput-optimal wireless link scheduling. In *Proceedings IEEE INFOCOM 2009*, pages 2331–2339, April 2009.

[9] F. Kelly, A. Maulloo, and D. Tan. Rate control in communication networks: Shadow prices, proportional fairness and stability. *Journal of the Operational Research Society*, 49:237–252, 1998.

[10] A. M. Khodaian and B. H. Khalaj. Delay-constrained utility maximisation in multi-hop random access networks. *Communications, IET*, 4(16):1908–1918, November 2010.

[11] M. Kodialam and T. Nandagopal. Characterizing the capacity region in multi-radio multi-channel wireless mesh networks. In *Proceedings MobiCom 2005*, pages 73–87, 2005.

[12] Long Bao Le, E. Modiano, and N. B. Shroff. Optimal control of wireless networks with finite buffers. In *Proceedings on IEEE INFOCOM*, pages 1–9, March 2010.

[13] R. Majumder, G. Bag, and Ki-Hyung Kim. Power sharing and control in distributed generation with wireless sensor networks. *IEEE Transactions on Smart Grid*, 3(2):618–634, June 2012.

[14] M. J. Neely, E. Modiano, and C. E. Rohrs. Dynamic power allocation and routing for time-varying wireless networks. *IEEE Journal on Selected Areas in Communications*, 23(1):89–103, January 2005.

[15] S. Oh, L. Schenato, P. Chen, and S. Sastry. A scalable real-time multiple-target tracking algorithm for sensor networks. *Memorandum*, 2005.

[16] S. Petersen and S. Carlsen. Wirelesshart versus ISA100.11a: The format war hits the factory floor. *IEEE Industrial Electronics Magazine*, 5(4):23–34, December 2011.

[17] H. L. Ren and M. Q.-H. Meng. Game-theoretic modeling of joint topology control and power scheduling for wireless heterogeneous sensor networks. *IEEE Transactions on Automation Science and Engineering*, 6(4):610–625, October 2009.

[18] S. Sakai, M. Togasaki, and K. Yamazaki. A note on greedy algorithms for the maximum weighted independent set problem. *Discrete Appl. Math.*, 126(2-3):313–322, 2003.

[19] I. Stoianov, L. Nachman, S. Madden, T. Tokmouline, and M. Csail. Pipenet: A wireless sensor network for pipeline monitoring. In *Proceedings 6th International Symposium on Information Processing in Sensor Networks*, pages 264–273, April 2007.

[20] P. Suriyachai, U. Roedig, and A. Scott. A survey of mac protocols for mission-critical applications in wireless sensor networks. *IEEE on Communications Surveys Tutorials*, 14(2):240–264, 2012.

[21] C. Taddia, G. Mazzini, M. K. Chahine, and K. Shahin. Reliable data forwarding for delay constraint wireless sensor networks. In *Proceedings 3rd International Conference on Information and Communication Technologies: From Theory to Applications*, pages 1–8, April 2008.

[22] Y. Takizawa, A. Yamaguchi, and S. Obana. Packet distribution for communications using multiple IEEE802.11 wireless interfaces and its impact on TCP. *IEICE Transaction on Communication*, E92-B(1):159–170, 2009.

[23] L. Tassiulas and A. Ephremides. Stability properties of constrained queueing systems and scheduling policies for maximum throughput in multihop radio networks. *IEEE Transactions on Automatic Control*, 37(12):1936–1948, December 1992.

[24] Y. Wang and S. Boyd. Performance bounds for linear stochastic control. *Systems, Control Letters*, 58(3).

[25] H. Zhang, H. Ma, X. Li, and S. Tang. In-network estimation with delay constraints in wireless sensor networks. *IEEE Transactions on Parallel and Distributed Systems*, PP(99):1, 2012.

5

Distributed Scheduling for Cooperative Tracking in Hierarchical Wireless Sensor Networks

Ying-Chih Chen, Su-Mong Yu and Chih-Yu Wen

Department of Electrical Engineering, Graduate Institute of Communication Engineering, National Chung Hsing University, Taichung 402, Taiwan
e-mail: cwen@dragon.nchu.edu.tw

Abstract

This chapter proposes a distributed method for cooperative target tracking in hierarchical wireless sensor networks. The concept of leader-based information processing is conducted to achieve object positioning, considering a cluster-based network topology. Random timers and local information are applied to adaptively select a sub-cluster for the localization task. The proposed energy-efficient tracking algorithm allows each sub-cluster member to locally estimate the target position with a Bayesian filtering framework and further performs estimation fusion in the leader node with the covariance intersection algorithm. This chapter evaluates the merits and trade-offs of the protocol design towards developing more efficient and practical algorithms for object position estimation.

Keywords: wireless sensor networks, target tracking, Bayesian filtering, covariance intersection.

Fabrice Theoleyre and Ai-Chun Pang (Eds.), Internet of Things and
M2M Communications, 93–114.

5.1 Introduction

Giving the limited power and processing capability in a sensor mote, a critical challenge of target tracking is how to acquire suitable data and perform information processing tasks at the local level through cooperative communication and networking in the vicinity of the target [1]. Thus, scalability and the need to conserve energy lead to the idea of hierarchically organizing the sensors, which can represent the target state and incorporate statistical models for the sensing schedule and target positioning. This chapter aims to develop a fully distributed method for cooperative target tracking in wireless sensor networks from two perspectives: (1) energy-balanced tracking and (2) improving estimation accuracy.

The first perspective is to build up an energy-balanced tracking network architecture. In this work, the concept of leader-based information processing is conducted to automatically achieve cooperative sensor scheduling with multiple tasking sensors in a cluster-based network topology based on sensor residual energy level, target information, and estimation quality. To avoid the ambiguity, a clusterhead and cluster members are referring to the original network topology. A leader and sub-cluster members are referring to the sensor group for the tracking task. Random timers and local criterions are used to determine the tracking responsibility of the clusters. Afterwards, a sub-cluster of the corresponding cluster for the tracking task is formed by a leader, which can be a clusterhead or a cluster member in the original cluster-based network. The second perspective is to explore the behaviors/characteristics of a target such that supplementary information can be applied to improve estimation accuracy. Within the sub-cluster, the sensing nodes provide their measurements to the leader. Upon receiving the measurements, the leader fuse the local estimates from the sub-cluster members and reports it to the clusterhead. When the target moves out the region of the current active sub-cluster, the leader needs to trigger the leader handoff procedure (detailed in Section 5.3.4).

As shown in [2], compared with the dynamic clustering approach in a flat network topology, the static clustering approach incurs a larger location error since a clusterhead may not be a good local controller for estimating the location and reporting the event due to target movement. However, given a fixed hierarchical network topology, dynamic clustering approaches may not be feasible. Therefore, considering a cluster-based network topology, the proposed algorithm aims to improve the energy efficiency and provide good estimation accuracy. Accordingly, the proposed tracking approach organizes a sub-cluster for the tracking task, allows each sub-cluster member to locally compute the target position, and uses cooperation to obtain the fused estimate in the leader node. The local level estimate of a sub-cluster member and the global level estimate of a leader can be derived by a Bayesian framework [3] and the covariance intersection algorithm [4], respectively.

The rest of the chapter is organized as follows. In Section 5.2 we briefly review the literatures on target tracking. In Section 5.3 we formulate the problem and derive a distributed solution for target positioning. Then, in Section 5.4 we consider the energy consumption of the proposed tracking scheme. In Section 5.5, the feasibility of the proposed scheme is examined via simulation. The performance comparison of the proposed approach and the scheme with a hierarchical network topology in [5] is presented. In Section 5.6 we discuss open research challenges toward effective sensor tasking and control. In Section 5.7 we present our conclusion and show future research directions.

5.2 Literature Review

There are five major categories for the target tracking solutions [2]: Tree-based Tracking, Cluster-based Tracking, Prediction-based Tracking, Mobicast Message-based Tracking, and Hybrid Tracking. Studies have shown that the cluster-based tracking algorithms have better network scalability and resource utilization compared with those in other categories. Prediction-based tracking rely on tree-based and cluster-based tracking in addition to prediction method, but the tracking accuracy cannot be guaranteed. Mobicast message-based tracking method depends on prediction, which is a multi-cast method in which message is delivered to a group of nodes that change with time according to estimated velocity of moving entity. Scheduling strategies vary in target tracking protocols and time synchronization may be needed to set the wake up and sleep timings of sensor nodes.

Since the proposed approaches fall into the category of cluster-based tracking, we focus on the research results of this category. The following subsections briefly describe the current literature of target tracking with respect to the number of tasking sensors.

5.2.1 Single Tasking Sensor

In the current literature, the general problem formation of target tracking is reformed to be a sensor selection problem with the uniform sampling interval and without incorporating the target dynamics. For instance, the information-driven sensor query (IDSQ) approach [6] and the entropy approach based on sensor selection [7]. In contrast, the authors in [8, 9] propose adaptive scheduling strategies to choose the single tasking sensor and determining the sampling interval simultaneously. In [8], the sensors are scheduled in two tracking modes, i.e., the fast tracking approaching mode when the predicted tracking accuracy is not satisfactory, and the tracking maintenance mode when the predicted tracking accuracy is satisfactory. The approach employs an Extended Kalman Filter (EKF) based estimation technique to predict the tracking accuracy, and adopts a linear energy model to predict the energy consumption.

In [9], the proposed algorithm applies the interactive multiple model (IMM) filter filter to estimate and predict the target's dynamic state and select the tasking sensor node and sampling interval for each time step based on both of the tracking accuracy and the energy cost. Simulation results show that the proposed approach outperforms the popular extended Kalman filter (EKF) based tracking scheme for maneuvering target in terms of tracking accuracy and energy efficiency. In [10], a small region is specified in order to select the single tasking sensor for achieving energy conservation. The distributed IMM filter is employed to estimate target position and velocity. A novel dynamic grouping idea is proposed to schedule next tasking node.

5.2.2 Multiple Tasking Sensors

For the purpose of increasing estimation accuracy and reliability, multiple tasking sensors may be scheduled to track the target with detection uncertainties. In [5], target localization strategies based on a communication protocol between the clusterhead and cluster members are presented. In these approaches, a subset of sensor nodes are selected for querying detailed target information. Although a certain amount of energy and communication bandwidth are conserved, the processing burden all falls in the clusterhead, which may drain its power quickly. Suganya [11] focuses on tracking error and energy management involved in tracking the target. In this approach, the sensor nodes collectively monitor and track the movement of the target, which involves detecting, clustering and localization of target.

In [12], an information-driven approach in ad hoc sensor networks is proposed to determine the tasking sensors in a "sensor collaboration" by dynamically optimizing the information utility of data for a given cost of communication and computation. In [13], a distributed estimation method is proposed based on mobile agent (MA) computing paradigm and generic sequential Bayesian filtering for the target state estimation at each time step. Nonetheless, the MA migration planning problem needs to be handled in order to achieve the desired tracking accuracy. The tracking schemes in [14, 15] combine the mechanisms of the tree-based and cluster-based schemes and propose information-based target tracking methods. However, the proposed sensor systems still have to deal with complexity issues. The authors in [16,17] propose multi-sensor scheduling schemes for maneuvering target tracking in sensor networks, while not considering the motion information of the target. Comprehensive surveys of design challenges and recently proposed target tracking algorithms can be found in [2].

Note that most of these design approaches are dynamic clustering protocols in a flat network. In contrast, the method in [5] is built upon a static cluster-based network topology. Thus, [5] may provide a good way to benchmark the performance of the proposed tracking scheme.

5.3 Distributed Target Tracking System

In this section, we present the proposed distributed cooperative target tracking system: *Two-level Clustering Approach via Timer* (TCAT) in a cluster-based network topology. Here level-one clustering indicates the original network topology with control of clusterheads. Level-two clustering means the sub-cluster formation for the tracking task with control of tasking leaders. Therefore, the information flow goes through the sub-cluster members to the leader, and then to the clusterhead, and vice versa. The main assumptions are (i) all sensors are homogeneous, (ii) the sensors are in fixed and known location, (iii) a pre-specified sub-cluster size n is applied to perform cooperative target positioning with angle-of-arrival information, (iv) the target periodically broadcasts a message for measurement purpose. Note that these assumptions may be applied to health-care scenarios or habitat monitoring to locate patients or animals. The distributed tracking architecture in a cluster-based network is depicted in Figure 5.1.

In Phases I and II, random timers and local information are applied to adaptively select a tasking leader and sub-cluster members for the localization task. In Phase III, the Bayesian particle filter [3, 18] is used to estimate the unknown target position from state equations. The objective is to find feasible position to make the error of state vector minimum. Afterwards, the covariance intersection algorithm [4] is adopted to perform estimation fusion.

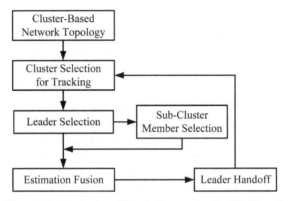

Figure 5.1 Illustration of block diagram for the TCAT method.

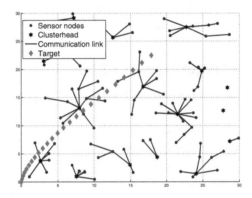

Figure 5.2 The cluster-based network topology and target movement with 25 time steps.

In order to maintain tracking stability, Phase IV performs the leader handoff task.

5.3.1 Phase I: Tasking Leader Selection

When sensors are first deployed, they may apply the CAWT [19] to establish the cluster-based network architecture (Figure 5.2). However, due to the target movement, the clusterhead may be not a proper local controller in the neighborhood of the target. Thus, a cluster member may be a good control candidate and can be a leader for the tracking task. Denote a sensor with tracking responsibility as an active sensor. Otherwise, a sensor is marked as an inactive sensor. Suppose each sensor is an inactive sensor with the initial deployment. The tracking task is triggered when the target broadcasts a message of L_{id}, where L_{id} is a leader ID with an initial value zero, which is used to inform the active sensors to compete for being a leader of the tracking task. Thus, when sensor i receives the message sent from the target, it will broadcasts a *Hello* message and become an active sensor, which allows itself to estimate how many neighboring active sensors it has. A *Hello* message consists of: (1) the sensor ID of the sending sensor, (2) the leader ID of the sending sensor, and (3) the cluster ID of the sending sensor. Therefore, the sensors update their neighboring information (i.e. a counter specifying how many neighboring active sensors it has detected) and decrease the random leader waiting time (LWT) through the received *Hello* message sent from neighboring active sensors.

Assume the initial value of the waiting time of sensor i, $LWT_i^{(0)}$, is a sample from the distribution $U(C, D')$, where $D' = C + D$, C and D are positive numbers, and $U(\cdot, \cdot)$ is a uniform distribution. The update formula for the random LWT is given by

$$LWT_i^{(j+1)} = \alpha \cdot LWT_i^{(j)}, \tag{5.1}$$

where $LWT_i^{(j)}$ is the random LWT of sensor i at time step j, and $0 < \alpha < 1$. Note that the setting of random LWT may depend on sensor residual energy level, target information, and measurement quality (e.g. the channel condition, the accuracy of positioning system). When the timer of sensor i expires, it then broadcasts a *Leader* message to claim that it is leader i (e.g. $L_{id} = i$) for the tracking task.

5.3.2 Phase II: Choosing the Sub-Cluster Members

Based on the claimed message sent from the leader and the cluster ID of the leader, the target will send a message to inform the active sensors about the corresponding cluster for the tracking task, which also notifies the active sensors with the same cluster ID to be the candidate sub-cluster members associated with the leader. To choose the members associated with a leader, instead of directly selecting the active sensors from the leader, the sensor selection may be determined based on the reporting order of target position estimates from the neighboring active sensors of the leader. Accordingly, a candidate sub-cluster member, say sensor m, may decrease its LWT along with an extra backoff time BT_m, which is inversely proportional to the estimation accuracy, for reporting the estimate of target position. When the timer

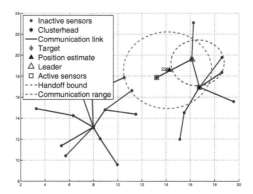

Figure 5.3 An example of leader and sub-cluster member selection.

LWT_m expires, sensor m will deliver the tracking information to leader i. That is, based on the time stamps of the received estimates, the target tracking group is then automatically formed with the leader.

For those active sensors without receiving a *Leader* message, they transmit the estimated target position directly to the clusterhead and become supplementary sub-cluster members. This is attributed to the fact that the active sensors may not have direct communication with the leader. Hence, they may send their tracking estimates to the clusterhead for providing supplementary information on the tracking task. Therefore, when the number of sub-cluster members meets the desired number n, the leader will perform the CI model (as detailed in Section 5.3.3.1) to obtain a global target position estimate and send a *Position* message to the clusterhead, which also serves to specify the final sub-cluster members for cooperative target tracking. If the leader does not collect sufficient number of estimates (i.e. $|M| < n$, M is the index set of the sub-cluster), then the leader may send its fused estimate to the clusterhead and request the clusterhead to incorporate the supplementary estimates if possible. Afterwards, the clusterhead will perform the CI algorithm based on the received fused estimate and the supplementary estimates. Note that in order to spread the energy burden in the network, the cluster is responsible for informing the base station about the target tracking and positioning. Figure 5.3 shows an example of leader and sub-cluster member selection.

5.3.3 Phase III: Target Positioning

Referring to [20], evaluating the computation process and the significance of approximate accuracy is an important step in deriving either exact or approximate solutions for the localization problem. This phase presents a measurement mechanism to assess the achievable estimation accuracy.

5.3.3.1 Covariance Intersection (CI)

Note that the tasking sensors perform target positioning with Bayesian particle filtering [3]. One of the main advantages of this approach is that the sensor carries along a complete distribution of estimates of target position. Therefore, the distribution is inherently a measure of the accuracy of the positioning system. For obtaining a global estimates, we adopt covariance intersection to perform data fusion. The CI method of [4] provides the best estimate given the information available, which takes a convex combination of mean and covariance estimates that are represented in information space. Since these typical runs are independent, the general form is

$$P_{cc}^{-1} = \omega_1 P_{a_1 a_1}^{-1} + \cdots + \omega_n P_{a_n a_n}^{-1}, \tag{5.2}$$

$$P_{cc}^{-1} c = \omega_1 P_{a_1 a_1}^{-1} a_1 + \cdots + \omega_n P_{a_n a_n}^{-1} a_n. \tag{5.3}$$

where $\sum_{i=1}^{n} \omega_i = 1$, $n > 1$, a_i is the estimate of the mean from available information, $P_{a_i a_i}$ is the estimate of the variance from available information, c is the new estimate of the mean, and P_{cc} is the new estimate of the variance.

5.3.3.2 Estimation Fusion

The distributed scheme is executed in two steps. (1) Group Estimation: local decisions are performed. (2) Estimation Fusion: a fusion rule is applied to combine the posterior density of the estimation from each member of the cooperative group in the leader sensor. Since the weight reflects the significance attached to the estimate, the next issue is to determine the weigh ω_i for each estimate and try to weight out faulty estimates. One strategy for choosing ω_i is to use the utility measure. Since the utility of a sensor measurement is a function of the geometric location of the target, here we consider the Mahalanobis measure [1]. Hence, with respect to a neighboring system estimate characterized by the mean $\mu_{m\ell}$ and covariance Σ, the utility function for sensor m is defined as the geometric measure

$$\mathcal{U}_{m\ell} = (\mu_{m0} - \mu_{m\ell})^T \Sigma^{-1} (\mu_{m0} - \mu_{m\ell}), \tag{5.4}$$

where μ_{m0} is the local estimated target position of sensor m and ℓ refers to a neighboring system estimate. In order to arrive at a consensus, the utility measure $\mathcal{U}_{m\ell}$ can be shown to be $\mathcal{U}_{m\ell} \leq 1$. Given the utility measure, two estimates can be allowed to be compared in a common framework and measure how much they differ $|\mu_{m0} - \mu_{m\ell}|$. Accordingly, the weights for the CI method in (5.2) may be determined by

$$\omega_\ell = \frac{\frac{1}{\mathcal{U}_{m\ell}}}{\sum_{k \in U_s} \frac{1}{\mathcal{U}_{mk}}}, \tag{5.5}$$

where U_s is the index set of the neighboring estimates that pass the utility test. Otherwise, ω_ℓ is set to be zero. Notice that in this work m may refer to a tasking leader and ℓ may refer to a sub-cluster member.

5.3.4 Phase IV: Leader Handoff

This phase performs the leader handoff task, which aims to maintain tracking stability. The conditions for initiating the handoff procedure are:

- The distance between the reference location of the sub-cluster member P_i ($\forall i \in M$) and the fused target position estimate $f(j)$ exceeds the handoff threshold value at time step j. That is, $d(P_i, f(j)) > \mathfrak{R}, \forall i \in M$.
- Due to the movement of the target, the number of *Position* messages or expected active sensors are less than the desired value. That is, $N_P < n$ or $N_E < n$.

For condition 1, $\mathfrak{R} = \beta \cdot R$, where R is the radio transmission range with $0 < \beta < 1$. For condition 2, a threshold value $\triangle R$ for dynamically measuring the number of valid sub-cluster members is applied. Figure 5.4 (top) presents an example of adaptively updating the sub-cluster members with $\triangle R$. Denote N_E as the number of active sensors which satisfy $d(P_i, B_{L_{id}}) \leq \triangle R$ with $\triangle R = \delta \cdot R$ ($0 < \delta < 1$), where leader handoff boundary $B_{L_{id}}$ is derived by the estimated target position, the positions of the sub-cluster members, and the bounding-box algorithm [21]. Denote N_P as the number of estimates from the sub-cluster members. If any of the above conditions holds, the leader and target will sequentially broadcast a *Handoff* message with $L_{id} = 0$ to trigger a leader reselection process as depicted in Phases I and II. Referring to Figure 5.2 (left), Figure 5.4 (bottom) shows an example of leader handoff procedure from time step 23 to time step 24.

Table 5.1 Procedures of the TCAT model for target tracking.

1. Target broadcasts a message with $L_{id} = 0$.
2. Determine the active sensors $I_A^{(j)}$ and the leader L_{id} at time step j.
    (a) $L_{id} = \text{argmin}_i LWT_i^{(j)}, i \in I_A^{(j)}; M = L_{id}$.
    (b) $BT_i = LWT_i + bf_i, i \in I_{CA}; I_{CA} = I_A^{(j)} \cap C_{L_{id}}$.
3. Find the sub-cluster members:
    while($I_{CA} \neq \emptyset$)
        (i) if$((S = I_{N_{L_{id}}}) == \emptyset), S = (I_{N_{L_{id}}} \cap I_{CA})^c$.
        (ii) $\widehat{M} = \text{argmin}_i BT_i^{(j)}, i \in S$.
        (iii) Send the position estimate to the leader or clusterhead.
        (iv) $M = \widehat{M}; S = (S \cap M)^c; I_{CA} = (I_{CA} \cap \widehat{M})^c$.
        (v) if$(|M| == n)$, break.
    end
4. Estimation Fusion:
    (a) Leader sends the global fused estimate $f(j)$ to the clusterhead.
    (b) The clusterhead disseminates the $f(j)$ to the base station.
5. Perform leader handoff (renewing sub-cluster members):
    (a) $N_P = |H_P|, H_P = \{i : d(P_i, f(j)) \leq R, i \in M\}$.
    (b) if$(N_P < n), j = j + 1$ and go to Step 1.
    (c) if$(d(B_{L_{id}}, f(j)) > \Re)$
        $M = (M \cap L_{id})^c$,
        $K = \{i : \text{argmin}_i d(P_i, f(i)) \leq \Re, i \in C_{L_{id}}\}$,
        $L_{id} = \text{argmax}_k N_s(k), k \in K$,
        if$(L_{id} == \emptyset), j = j + 1$; go to Step 1.
    end
    (d) $I_{CA} = (C_{L_{id}} \cap M)^c$.
    (e) $N_E = |H_E|, H_E = \{i : d(P_i, B_{L_{id}}) \leq \Delta R, i \in M\}$.
    (f) while($N_E < n$)
        (i) $\widehat{M} = \text{argmin}_i d(P_i, B_{L_{id}}) \leq R, i \in I_{CA}$.
        (ii) $N_E = N_E + 1; M = M \cup \widehat{M}; I_{CA} = (I_{CA} \cap \widehat{M})^c$.
        (iii) if$(I_{CA} == \emptyset)$, break.
    end
    (g) if$(N_E < n), j = j + 1$; go to Step 1.
    (h) Target broadcasts a message with L_{id} at time step $j + 1$.
    (i) Go to Step 4.

Due to the cluster-based network topology, the handoff schemes can be further divided into two categories: (1) intra-cluster leader handoff and (2) inter-cluster leader handoff. Since the clusterhead collects the supplementary estimates and receives the estimate from the leader, it may monitor the $d(P_i, f(j))$ and N_E. If condition 1 holds, then an intra-cluster leader handoff is performed and the clusterhead may assign an cluster member to be a new

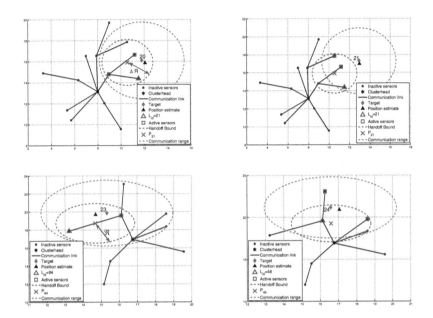

Figure 5.4 An example of adaptively updating the sub-cluster members with $\triangle R$ (top); An example of leader hand off procedure (bottom).

leader. Otherwise, an inter-cluster leader handoff is triggered and go to the operations in Phase I and II. The procedures of the TCAT model for cooperative target tracking are detailed in Table 5.1. Note that I_{N_i} is the index set of vicinal sensors of sensor i, $N_s(i)$ is the number of neighboring sensors for sensor i, and C_i is the set of cluster members of sensor i.

5.4 Analysis of Energy Consumption

This section considers the energy consumption of the proposed scheme in Section 5.3. The total power requirements include both the power required to transmit messages and the power required to receive (or process) messages. Suppose that the energy needed to transmit for sensors with omnidirectional antennas is E_T, which depends on the transmitting range R, and the energy needed to receive is E_R.

When the target broadcasts a message with $L_{id} = 0$, its neighboring sensor, say sensor i, becomes an active sensor and broadcasts a *Hello* message with a random waiting time $LWT_i^{(j)}$ for being a task leader at time step j. As the active sensor i claims to be a leader, the L_{id} is updated and broadcasted from target. As a result, the number of transmissions and receptions for tasking leader selection (Phase I) are

$$S_T^c(j) = 1 + N_t(j) \tag{5.6}$$

$$S_R^c(j) = \sum_{i \in (I_A^{(j)} \cup L_{id}^{(j)})} |I_{N_i}| + 2N_t(j) \tag{5.7}$$

where $L_{id}^{(j)}$ is a leader ID at time step j, I_{N_i} is the index set of vicinal sensors of sensor i, and $N_t(j)$ is the number of transmissions of vicinal sensors of the target at time step j.

Afterward, the active sub-cluster members of the leader are selected according to the extra backoff time BT_m, which transmit the *Position* messages to the leader. Thus, the number of transmissions yields the sub-cluster size $|M(j) \cap I_{N_{L_{id}^{(j)}}}|$ and the number of receptions is $|I_{N_i}|(\forall i \in M(j) \cap I_{N_{L_{id}^{(j)}}})$, where the $M(j)$ is the set of sub-cluster members at time step j. Since the fused estimate $f(j)$ is routed to the clusterhead in a multi-hop manner, the number of transmissions is $N_i^H(\forall i \in (I_A^{(j)} \cap I_{N_{L_{id}^{(j)}}})^c)$, where N_i^H is number of hops for sensor i to perform estimation reporting. Finally, the clusterhead disseminates $f(j)$ to the base station. Furthermore, if the clusterhead receives a message from the leader for incorporating supplementary estimates, it may assign a desired number of supplementary members to be supplementary sub-cluster members for the tracking task. Therefore, the number of transmissions and receptions for selecting tasking members (Phase II) and delivering the position estimate (Phase III) are

$$S_T^p(j) = \left| M(j) \cap I_{N_{L_{id}^{(j)}}} \right| + \sum_{i \in (I_A^{(j)} \cap I_{N_{L_{id}^{(j)}}})^c} N_i^H + \sum_{i \in M'} N_i^H + 1 \tag{5.8}$$

$$S_R^p(j) = \sum_{i \in (M(j) \cap I_{N_{L_{id}^{(j)}}})} |I_{N_i}| + \left(\sum_{i \in (I_A^{(j)} \cap I_{N_{L_{id}^{(j)}}})^c} + \sum_{i \in M'} \right) \sum_{h=1}^{N_i^H} N_{I_{R_i}^{(h)}} \tag{5.9}$$

where $M' = \{i : \text{argmin}_i \sum N_i^H, i \in M(j) \cap (I_A^{(j)} \cap (I_{N_{L_{id}^{(j)}}} \cup L_{id}^{(j)}))^c\}$ and $I_{R_i}^{(h)}$ is index set of relay ID for sensor i at hop h.

As characterized in Phase IV, since the clusterhead has the capability of updating the sub-cluster members, a cluster member, say sensor i, may become an active supplementary sub-cluster member or leader when receiving the message, which contains the IDs of new supplementary sub-cluster member or the leader (i.e. $(i \cap \overline{M}(j) \neq \emptyset)$ or leader ID), and then joint the tracking task. Note that $\overline{M}(j)$ is the index set of new supplementary sub-cluster members. Therefore, we obtain

$$S_T^f(j) = \sum_{i \in M''} N_i^H + 1 \tag{5.10}$$

$$S_R^f(j) = N_{CH(j)} + \sum_{i \in M''} \sum_{h=1}^{N_i^H} N_{I_{R_i}^{(h)}} \tag{5.11}$$

where $M'' = \{i : \text{argmin}_i \sum N_i^H, i \in (\overline{M}(j) \cap I_{NCH(j)})^c \cup (L_{id}^{(j+1)} \cap L_{id}^{(j)})^c\}$ and $S_T^f(j)$ and $S_R^f(j)$ are the number of transmissions and receptions for leader handoff, respectively. Nonetheless, when the leader ID is zero (i.e. $L_{id}^{(j+1)} = 0$), the sub-cluster members are assigned to be inactive sensors and the procedure of selecting a new sub-cluster will be triggered. Accordingly, the total energy consumption of transmission and reception for tracking target is $E^{TCAT} = \sum_{j=1}(E_T \cdot (S_T^c(j) + S_T^p(j) + S_T^f(j)) + E_R \cdot (S_R^c(j) + S_R^p(j) + S_R^f(j)))$.

5.5 Simulation

To evaluate the performance of the proposed approach, assume that the target moves within the $x - y$ sensing field according to the standard second-order model [3]

$$X_k = \Phi X_{k-1} + \Gamma \mathbf{w}_k \tag{5.12}$$

over a four-dimensional state space, where $X_k = (x, \dot{x}, y, \dot{y})_k^T$, $\mathbf{w}_k = (w_x, w_y)_k^T$, an uncorrelated Gaussian diffusion term describing the uncertainty,

$$\Phi = \begin{pmatrix} 1 & 1 & 0 & 0 \\ 0 & 1 & 0 & 0 \\ 0 & 0 & 1 & 1 \\ 0 & 0 & 0 & 1 \end{pmatrix}, \text{ and } \Gamma = \begin{pmatrix} 0.5 & 0 \\ 1 & 0 \\ 0 & 0.5 \\ 0 & 1 \end{pmatrix}.$$

Here x and y denote the Cartesian coordinate of the target. The noisy measurement is given by

$$z_k = \tan^{-1}(y_k/x_k) + v_k, \tag{5.13}$$

where the measurement noise, v_k, is a zero mean Gaussian white noise process with a finite variance σ_θ^2. Before measurements are taken at $k = 1$, the initial state vector is assumed to be a Gaussian distribution with known mean $\bar{x}_1$, and covariance

$$M_1 = \begin{pmatrix} \sigma_1^2 & 0 & 0 & 0 \\ 0 & \sigma_2^2 & 0 & 0 \\ 0 & 0 & \sigma_3^2 & 0 \\ 0 & 0 & 0 & \sigma_4^2 \end{pmatrix}.$$

The target trajectory and measurements are generated based on equations (5.12) and (5.13) with the parameter values: the covariance of the system noise, $Q = qI_2$, where I_2 is the 2 x 2 identity matrix, $\sqrt{q} = 0.001$. The initial state of the target is $x_1 = (0.0, 0.1, 0.0, 0.05)^T$. The prior distribution parameters are set to $\bar{x}_1 = (0.0, 0.0, 0.4, 0.05)^T$ and $\sigma_1 = 0.5$, $\sigma_2 = 0.001$, $\sigma_3 = 0.05$ and $\sigma_1 = 0.01$. Assume the target moves continuously and applies the dynamics in [3]. The particle filtering with $N_{PF} = 1000$ samples and the CI method are used to obtain target position estimate for 25 time steps.

Figure 5.5 depicts the system performance (e.g. the average positioning error and the leader handoff frequency) with various values of parameters (α, β, δ, sub-cluster size, and network density), for cooperative target tracking. Observe that for parameter β such as Figure 5.6, there is a tradeoff between localization error and leader handoff frequency since a larger value of β (i.e. a larger handoff threshold value) may lead to a lower leader handoff frequency and may result in a lower positioning accuracy. Without loss of generality, we investigate the typical performance of the TCAT in a network with random uniform deployment of N_S sensors given $\alpha = \beta = \delta = 0.5$, $C = D = 1$, and the standard deviation of angle information $\sigma_\theta = 0.5$ radian. Note that the entire experiments are conducted in a square with side length $L = 30$ unit length and transmission range $R = L\sqrt{\log_{10}(L)/N_S}$ [22].

5.5.1 Target Localization Error

To assess the tracking accuracy, the root mean square error is used for comparing the tracking accuracy of the distributed TCAT with that of [5]. Referring to the network topology and the target movement in Figure 5.2, we

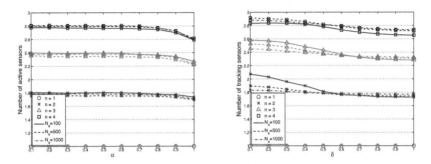

Figure 5.5 The target localization error with $\beta = \delta = 0.5$ and varying the values of α (left); The target localization error with $\alpha = \beta = 0.5$ and varying the values of δ (right).

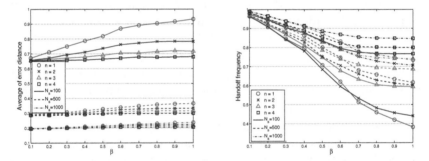

Figure 5.6 The target localization error with $\alpha = \delta = 0.5$ and varying the values of β (left); The leader handoff frequency with $\alpha = \delta = 0.5$ and varying the values of β (right).

vary the number of sub-cluster members from 1 to 4. Figure 5.7 (left) shows the accuracy of the position estimate. The performance improves along with the number of sub-cluster size n. However, the improvement is not significant (especially when the number n is greater than 2). This suggests that even a low number of sub-cluster members can also achieve good estimation accuracy.

As illustrated in Figure 5.7 (left), the radio transmission range R is assumed to be the same as the target detection range R_e. Here the effect of varying target detection ranges on the performance is investigated with changing the ratio of the radio transmission range to target detection range. Figure 5.7 (right) depicts that when the ratio is greater than one (i.e. $R/R_e > 1$), the sensors may fail to detect most target events and a larger net-

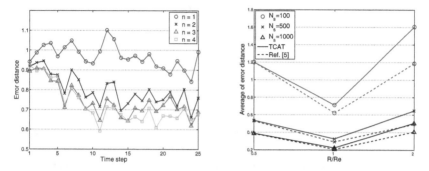

Figure 5.7 The target localization error (left); The estimation error with $n = 2$ and various ratios of R/R_e (right).

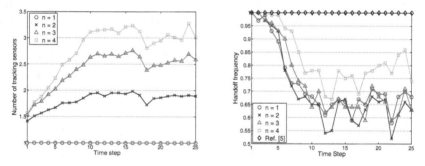

Figure 5.8 The typical runs of sub-cluster formation (left); the frequency of leader handoff (right).

work density may be required to detect the target of small signal magnitude and to suppress the estimation error. However, the estimation error decreases dramatically when the ratio $R/R_e \leq 1$ due to sufficient detection coverage.

5.5.2 Protocol Characteristics

Figure 5.8 (left) shows the typical runs of sub-cluster formation with $N_S = 100$ and $R/R_e = 1$. Notice that the TCAT effectively organizes the sensors into tracking groups. Referring to Figures 5.7 (left) and 5.8 (right), observe that compared with the protocol in [5], the TCAT has a lower leader handoff frequency and there is no significant performance degradation during

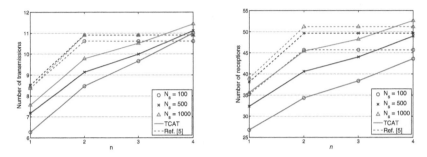

Figure 5.9 The comparison of accumulated network energy consumption.

the leader handoff period. Moreover, considering different sub-cluster size n and a network with random uniform deployment, Figure 5.7 suggests that compared with the TCAT with $n = 2$, the TCAT with a smaller value of n (e.g. $n = 1$) with respect to handoff condition 1 and the TCAT with a larger value of n (e.g. $n = 3, 4$) with respect to handoff condition 2 lead to a higher frequency of inter-cluster handoff. Thus, the TCAT with $n = 2$ may provide flexibility and robustness for distributed sensor scheduling management.

5.5.3 Network Energy Consumption

The simulation is performed with different density of nodes, considering the number of messages transmitted and received involved in clustering and target localization. Figures 5.9 show the accumulated energy consumption comparison between the TCAT scheme and the method in [5]. Observe that the number of transmissions/receptions grows nearly linearly as the tracking sub-cluster size increases. Referring to Figure 5.9, compared with a network with $N_S = 100$, a network with a larger scale (e.g. $N_S = 500, 1000$) may have a larger cluster size, which may lead to a higher energy consumption for data gathering at each round and result in a faster network resource depletion. Nonetheless, the performance of TCAT is still superior to that of the approach in [5].

Given $n = 2$, $N_S = 100$, and $R/R_e = 1$, Figure 5.7 (right) shows that the tracking accuracy of TCAT is comparable to that of [5]. Moreover, as depicted in Figure 5.9 (left), the number of transmissions with TCAT is about 19% less than that of Zou and Chakrabarty [5] and in Figure 5.9 (right) the number of receptions with TCAT is about 26% less than that of Zou and

Chakrabarty [5], which implies that the scheme in [5] may lead to a fast network energy depletion. Accordingly, the TCAT provides better network service characteristics compared to the protocol in [5].

Observe that in Figure 5.7 (left), although the performance of TCAT with $n = 1$ leads to a lower network energy consumption, compared with those of TCAT with $n = 2 \sim 4$, it results in a larger target localization error. Moreover, due to high correlation of sensing data in time and spatial domains, the sub-cluster with $n > 2$ members may lead to undesired sensing redundancy. Therefore, considering the trade-off between performance and network energy consumption, the TCAT with $n = 2$ may be a good choice for the tracking task.

5.6 Open Challenges

Though many target tracking schemes have been developed, feasible wireless sensor-based tracking systems require more breakthroughs in terms of network architecture, system design, and data processing techniques. Open challenges for designers proposing tracking design ideas are:

- *The design principles with a thorough knowledge of the interactions between the design parameters and the target information*

  Currently, very few papers consider performing cooperative positioning with the knowledge of target mobility models. Since the target motion information can be regarded as a hint for sensor tasking and control, the system performance may be enhanced and the estimation accuracy may be improved. Considering the impact of mobility models on positioning accuracy, analyze and discuss about the robustness of the proposed algorithm under different mobility models of the target, such as random models (Random Walk with Wrapping, Random Waypoint with Steady State), mobility models with temporal dependency (Gauss Markov Mobility Model, Semi-Markov Smooth Mobility Model), mobility models with spatial dependency (Reference Point Group Model, Community Based Mobility Model), and mobility models with geographic restriction (Pathway Mobility Model).

- *The distributed designs that will have the most significant impact on network performance*

Due to the limited processing capability of sensor motes, developing low-complexity energy-efficient data processing techniques for wireless sensor-based tracking architectures is an important issue. We believe there are many ways to improve the performance of the proposed schemes:

- by performing the data processing to identify key features and obtain local estimates at the sensor level (e.g. compressed sensing).
- by quantifying the trade-offs between amount of communication, speed of computation, and the cluster formation.
- by examining alternate ways to 'incorporate' the received neighborhood data for grouping the sensors into clusters.
- by using a notion of the reliability of the received neighborhood data for selecting good clusterhead/leader candidates.
- by applying the information of neighborhood residual energy distribution for initializing effective sensor tasking and control.

- *The cooperative designs with a thorough knowledge of the interactions between the tasking sensors for stably performing the handoff procedure and the tracking task*

Since handoff latencies affect the service quality of real-time applications, new handoff procedures should be developed in order to improve the leader handoff latency. Thus, we plan to

- characterize all handoff delay components in terms of service disruption, the delay of time, and cost implications required during the handoff procedure.
- provide an analysis of the handoff procedure on communication complexity, time complexity, and computational complexity.

5.7 Conclusion

Because of the resource-constrained sensors, feasible wireless sensor-based tracking systems require more breakthroughs in terms of network architecture, system design, and data processing techniques. In order to achieve good tracking quality, the number of sensors chosen for target positioning may be dynamically adjusted based on the available target and sensor information. Thus, incorporating the target motion information into cooperative positioning schemes with multiple sensors may be a good strategy to improve the estimation accuracy. Future plans will involve generalizing the method to implement a prototype of the tracking system, evaluate the merits of different cooperative schemes, explore the characteristics of target mobility model, and further examine the impact of target motion information on cooperative estimation performance.

References

[1] F. Zhao and L. J. Guibas. *Wireless Sensor Networks: An Information Processing Approach*. Morgan Kaufmann, USA, 2004.

[2] S. Bhatti and J. Xu. Survey of target tracking protocols using wireless sensor network. In *Proc. of the 5th ICWMC*, 2009.

[3] N. J. Gordon, D. J. Salmond, and A. F. M. Smith. Novel approach to nonlinear/non-Gaussian Bayesian state estimation. *IEEE Proceedings F, Radar and Signal Processing*, 140(2):107–113, 1993.

[4] S. Julier and J. K. Uhlmann. *General decentralized data fusion with Covariance Intersection(CI)*. Handbook of multisensor data fusion, USA, 2001.

[5] Y. Zou and K. Chakrabarty. Target localization based on energy considerations in distributed sensor networks. *Ad Hoc Networks*, 1(2).

[6] H. Haussecker, M. Chu, and F. Zhao. Scalable information-driven sensor querying and routing for ad hoc heterogeneous sensor networks. *International Journal of High Performance Computing Applications*, 16(3):90–110, 2002.

[7] Kung Yao, G. Pottie, and D. Estrin. Entropy-based sensor selection heuristic for target localization. In *Proceedings of the 3rd International Symposium on Information Processing in Sensor Networks*, pages 36–45, 2004.

[8] Jianyong Lin Wendong Xiao, Sen Zhang, and Chen Khong Tham. Energy-efficient adaptive sensor scheduling for target tracking in wireless sensor networks. *Journal of Control Theory and Applications*, 8(1):86–92, 2010.

[9] S. Zhang, W. Xiao, M. H. Ang, et al. Imm filter based sensor scheduling for maneuvering target tracking in wireless sensor networks. In *Proceedings of the International Conference on Intelligent Sensors, Sensor Networks and Information*, pages 287–292, 2007.

[10] B. Xu, Y. Liu, and L. Feng. Distributed IMM filter based dynamic-group scheduling scheme for maneuvering target tracking in wireless sensor network. In *Proceedings of the 2nd International Congress on Image and Signal Processing*, pages 1–6, 2009.

[11] S. Suganya. A cluster-based approach for collaborative target tracking in wireless sensor networks. *Proc. of ICETET*, 1(2), 2008.

[12] J. Shin, F. Zhao, and J. Reich. Information-driven dynamic sensor collaboration for tracking applications. *IEEE Signal Processing Magazine*, 19(2):61–72, 2002.

[13] H. Liu, J. Wei, and Z. Yu. An energy-efficient target tracking framework in wireless sensor networks. *EURASIP Journal on Advances in Signal Processing*, 2009:1–14, 2009.

[14] X. Wu, G. Huang, D. Tang, and X. Qian. A novel adaptive target tracking algorithm in wireless sensor networks. In R. Chen (Ed.), *ICICIS 2011, Part II, CCIS 135*, pages 477–486, 2011.

[15] A. Oka and L. Lampe. Distributed target tracking using signal strength measurements by a wireless sensor network. *IEEE JSAC*, 28(7):1006–1015, 2010.

[16] Y. Liu, B. Xu, and L. Feng. Energy-balanced multiple-sensor collaborative scheduling for maneuvering target tracking in wireless sensor networks. *Control Theory and Applications*, 9(1):58–65, 2011.

[17] T. He, S. Krishnamurthy, J. A. Stankovic, T. Abdelzaher, L. Luo, R. Stoleru, T. Yan, L. Gu, G. Zhou, J. Hui, and B. Krogh. Vigilnet: An integrated sensor network system for energy-efficient surveillance. *ACM Transactions Sensor Networks*, 2(1):1–38, 2006.

[18] M. S. Arulampalam, S. Maskell, N. Gordon, and T. Clapp. A tutorial on particle filters for online nonlinear/non-Gaussian Bayesian tracking. *IEEE Trans. on Signal Processing*, 50:174–188, 2002.

[19] C.-Y. Wen and W. A. Sethares. Automatic decentralized clustering for wireless sensor networks. *EURASIP JWCN*, 2005(5):686–697, 2005.

[20] T. Eren, D. K. Goldenberg, A. S. Morse, W. Whiteley, Y. R. Yang, B. D. O. Anderson, and P. N. Belhumeur. A theory of network localization. *IEEE Trans. Mob. Comput*, 5(12):1663–1678, 2006.

[21] A. Savvides, H. Park, and Mani B. Srivastava. The bits and flops of the n-hop multilateration primitive for node localization problems. In *Proceedings ACM Int. Workshop (WSNA)*, pages 112–121, 2002.

[22] Paolo Santi. *Topology Control in Wireless Ad Hoc and Sensor Networks*. Wiley, UK, 2005.

6

Time Synchronization on Cognitive Radio Ad Hoc Networks: A Bio-Inspired Approach*

Michele Nogueira[1], Aravind Kailas[2], Nadine Pari[1] and
Bhargavi Chandrakumar[2]

[1]*Department of Informatics – NR2, Federal University of Paraná, Brazil*
[2]*Department of Electrical and Computer Engineering, University of North
Carolina, Charlotte, USA*
*e-mail: michele@inf.ufpr.br, aravindk@ieee.org, nelpari@inf.ufpr.br,
bchandr4@uncc.edu*

Abstract

Harnessing the full power of the paradigm-shifting cognitive radio ad hoc networks (CRAHN) hinges on solving the problem of time synchronization between the radios during the different stages of the cognitive radio cycle. The dynamic network topology, the temporal and spatial variations in spectrum availability, and the distributed multi-hop architecture of CRAHN mandate novel solutions to achieve time synchronization and efficiently support spectrum sensing, access, decision and mobility. In this paper, we advance this research agenda by proposing the novel Bio-inspired time SynChronization protocol for CRAHN (BSynC). The protocol draws on the spontaneous firefly synchronization observed in parts of Southeast Asia. The significance of BSynC lies in its capability of promoting symmetric time synchronization between pairs of network nodes independent of the network topology or a predefined sequence for synchronization. It enables the nodes in CRAHN to synchronize in a decentralized manner efficiently, and is

*This work was partially supported by a grant from the REUNI Brazilian program.

*Fabrice Theoleyre and Ai-Chun Pang (Eds.), Internet of Things and
M2M Communications,* 115–132.

reliable. The findings suggest that BSynC improves convergence time, thereby favoring deployment in dynamic network scenarios.

Keywords: time synchronization, cognitive radio networks, bio-inspired solutions, resilience.

6.1 Introduction

Cognitive radio ad hoc networks (CRAHN) are promising candidates for effective spectrum management in a system comprising licensed primary users (PUs) and distributed unlicensed secondary users (SUs) (also known as cognitive radios (CRs). The SUs communicate with one another in a multi-hop way by opportunistically accessing spectrum holes (portions of the licensed spectrum not being used by PUs for prolonged periods of time). However, leveraging the full potential of CRAHN depends on time synchronization (among SUs) during the different stages of the dynamic spectrum management, such as spectrum sensing, decision, sharing, and mobility [1].

Further, effective time synchronization assists in overcoming the artifacts of the wireless channel, such as shadow fading that impede user-coordination, and also in avoiding interference with the PUs. The asynchronous nature of the distributed SUs, especially in mobile conditions, where they move at different speeds and along different directions makes time and frequency synchronization very challenging [15]. SUs will have different values for transmission times, sample and carrier frequencies, and thus inducing offsets in transmission times and sample frequencies in non-coherent communication systems, and divergent carrier frequencies in coherent communications.

In this chapter, the novel Bio-inspired SynChronization (BSynC) protocol has been proposed for CRAHN, mainly aimed at mitigating the effects of the aforementioned issues. BSynC synchronizes pairs of nodes "on-the-fly" and when compared to one of the popular synchronization protocols for wireless ad hoc networks, the timing-sync protocol in sensor network (TPSN) [8], it outperforms the traditional sender-receiver based synchronization in terms of the speed of achieving network-wide synchronization, and resiliency to link disruptions owing to node mobility.

6.1.1 Inspiration from Biology

The dynamics of the biological systems comprise simple generic rules that tote up to the complex behavior of a group, which is termed as swarm intelligence (SI). But, where this intelligence comes from raises a fundamental question in nature. The recent advances in the area of SI suggest strategies that can be adapted to manage complex systems. In other words, mimicking biology to solve real-world problems in wireless communication consists of three steps: (i) Identifying the various biological system(s) and their counterparts in the wireless domain; (ii) formulating accurate models; and (iii) translating the model into the wireless world application. Bio-inspired techniques have been applied to solve key problems in communication technologies [3]. More specifically, the phenomenon of synchronization has been investigated in large biological systems [3].

6.1.2 Why Fireflies?

Fireflies provide one of the most spectacular examples of synchronization in nature, and is undoubtedly a fascinating form of SI. At night in certain parts of southeast Asia thousands of male fireflies of some species congregate in trees and flash in synchrony. Conceptually, when one firefly sees the flashing of another firefly in its neighborhood, it shifts its flashing rhythm in accordance to the neighboring fireflys' flashing rhythm. The firefly synchronization is adaptive to the changes in the environment offering scalability in a fully distributed fashion, thereby justifying our choice for this biological model in synchronizing pairs of SUs (nodes) irrespective of the network topology.

We assume a CRAHN comprising very few nodes with access to a global time reference such as the universal coordinated time (UTC), and this is not an unreasonable assumption. They will serve as the ad hoc "master" nodes for the other nodes by periodically broadcasting their *time stamps*. The other nodes infer their timing from these broadcasts. It is remarked that this assumption is also valid from a real-world implementation stand-point as enabling each node with global time reference capabilities will prove to be less efficient and more expensive. Upon receiving the broadcast from the master node, the neighboring nodes adjust their clocks to ensure that the offsets are minimized. This iterative process occurs throughout the network until the nodes are synchronized.

The performance of BSynC is assessed by simulations under static and mobile scenarios. Preliminary results show that the convergence time of BSynC is smaller than the convergence time of TPSN. Furthermore, our findings suggest that BSynC outperforms TPSN in dynamic scenarios, managing efficiently changes in the network topology caused by spectrum handoffs, failures and others. BSynC presented satisfactory results independent on the number of master nodes.

The rest of the chapter is organized as follows. Section 6.2 presents the related work. Section 6.3 describes the analytical model, inspired by the Mirollo-Strogatz (MS) model for firefly synchronization [12], as well as the assumptions and terminology used in the analysis. Section 6.4 describes the novel BSynC protocol, and Section 6.5 presents the performance evaluation of the protocol and our initial findings. Finally, conclusions and some directions for future research are presented in Section 6.6.

6.2 Related Works

The main objective of the spectrum sensing and spectrum management in CRAHN is for the SUs to opportunistically leverage the unused portions of the spectrum without interfering with the PUs [10, 13]. To facilitate this opportunistic spectrum sharing (OSS), the CRAHN must device a mechanism such that when the SU detects the presence of PUs on a wireless channel, it must immediately switch to an unused one. On the other hand, if the SU detects the presence of another SU, it should employ a coexistence coordination mechanism to share spectrum resources efficiently. In a multi-user environment, the detection of PUs and the coexistence of SUs pose a multitude of challenges, thereby hindering the performance of CRAHN drastically. This highlights the importance of synchronization that is needed among the SUs and PUs, a long-standing challenge, especially from the implementation view-point. With this in mind, we briefly review previous work, which are most closely related to our proposed protocol. First, we present some of the popular approaches for wireless sensor networks (WSN), and then discuss the relevant works specifically addressing CRAHN.

The state-of-the-art in network-wide time synchronization is based on identifying a common global time reference. Solutions to time synchronization in decentralized wireless networks, such as sensor networks and mobile ad hoc networks, have been studied by the authors of [22] (and the references within). For example, the reference-broadcast synchronization (RBS) leverages the inherent broadcast property in wireless networks to share the offset and rate difference for the nodes to adjust their clocks relative to their neighbors [4], the Tiny-Sync and Mini-Sync (TS-MS) in [19] uses two-way messaging to estimate and maintain the relative clock drifts and offsets. While the former has the disadvantages associated with physical broadcasting, the latter mandate high computation and memory making them unsuitable for

CRAHN scenarios. that A tree-based algorithm, the timing sync protocol for sensor networks (TPSN) achieves synchronization along the branches of a logical tree at the root of which is the parent node has been very popular for WSN [6]. A lesser complexity version, the lightweight tree-based synchronization (LTS) have been proposed for both centralized and decentralized networks [9]. One of the key differences being that instead of using a spanning tree, the nodes determine resynchronization period based on desired clock accuracy, distance to reference node, clock drift and time of last synchronization. Since these protocols were primarily designed for WSN, they do not take into account mobility aspects, adaptability, or fault tolerance priorities that are relevant for CRAHN. In general, perfect time synchronization in massively distributed systems is a complex issue and difficult to solve [21], and most of these solutions cannot be directly applied to CRAHN due to channel handoffs and their self-adaptive features.

Some researchers have observed that synchronicity is a useful abstraction in many contexts and applications [14, 18]. Also, few existing synchronization protocols have taken advantage of bio-inspired models [3, 21]. Those protocols have been inspired by the first biological experiments carried out by Richmond, that developed mathematical models of synchronization. Werner et al. present an algorithm for synchronous WSN, called the reach-back firefly algorithm (RFA) [21], and was implemented for TinyOS synchronicity. It is based on MS model and on a mathematical model that describes how neurons spontaneously synchronize. The RFA considers realistic effects in the communications networks of sensors. Some researchers have extended the MS model where the nodes agree on a relative time reference to address the issue of synchronization when two asynchronous sets of nodes merge [20].

To the best of our knowledge, cognitive radio (CR)-Sync is the only synchronization protocol designed to take into account cognitive radio networks characteristics [15]. CR-Sync is based on the aforementioned TPSN, and hence suffers from the drawbacks of tree-based algorithms. Numerous medium access control (MAC)-level scheduling and synchronization algorithms for channel access in CRAHN have been proposed – Distributed channel assignment (DCA)-based MAC [16], Cognitive medium access control (C-MAC) [2], Opportunistic cognitive MAC protocol [17], SYN-MAC [11], to name a few. However, maintaining time synchronization network wide in accordance with the dynamics of the system is the most basic, but an important open issue that is still not addressed with utmost accuracy. The need to address the challenges posed in the already existing protocols triggered us to turn to Mother Nature to solve the long standing issue of synchronization in CRAHN. Our contribution is the design of the BSynC protocol, proposed to combat frequent switches of used channel by SUs on CRAHN. Based on the MS model, BSynC achieves temporal synchronicity between nodes in a flexible, self-adaptive, and fault-tolerant way.

6.3 Preliminares

In this section, the analytical system models, assumptions, and the terminology that are essential to the treatment and understanding of the proposed BSynC protocol are presented.

6.3.1 Firefly Synchronization Model

CRAHN solutions mandate aspects such as decentralization, flexibility, and node autonomy, and the proposed protocol draws on spontaneous synchronization exhibited by fireflies in nature for its decentralized and autonomous self-organization. What follows is a brief description of the Mirollo-Storgatz represents as an oscillator behavior of the of the internal clock (in a firefly) dictating when to flash, and the phase of this oscillator is modified upon reception of an external flash. The model assumes devices equipped with pulse-coupled oscillators that have an oscillator time period, T, after which they emit pulses. Each device has an internal clock, t, and $0 \leq t \leq T$. When $t = T$, the device emits a pulse, thereby resetting t to 0. Upon reception of a pulse from other oscillators, this internal clock is adjusted, and over time, synchronization emerges, i.e., pulses of different oscillators are transmitted simultaneously. This self-synchronizing behavior across a network of these devices is analogous to fireflies flashing synchronization.

Initially, the different internal clocks of the devices (across the network) are not synchronized because they can start the synchronization procedure at different moments. Hence, when a device pulses, another device in its vicinity responds to this by slightly increasing the value of the internal clock, t. The amount of adjustment of t is determined by the pulse function, $f(t)$ and a parameter ϵ, being a constant less than 1. Assuming that a device observes the pulsing of its neighbor at time $t = t'$, then the device under consideration sets its internal clock $t = t''$, where t'' is given by Eq. (6.1). It is remarked that $f(t)$ depends on the characteristics of the network, thus it should be defined in order to represent them. For instance, in [21], the authors define $f(t)$ as $\ln(t)$ for achieving synchronization in WSN taking into account realistic radio

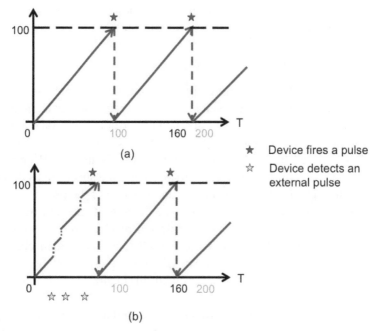

Figure 6.1 An illustration of the internal clock readjustment during the synchronization phase.

effects.

$$t'' = f^{-1}(f(t') + \epsilon) \tag{6.1}$$

The main parameter that affects the behavior of the system is ϵ, which determines the extent to which a node responds when it observes a neighbor pulse. A node responds to a neighbor pulse by incrementing its phase (shorting its time to pulse). Choosing a larger ϵ means that a node will take larger jumps in response to other nodes pulses, thus achieving synchrony faster. However, if ϵ is too large, then nodes can overshoot, preventing convergence. Making ϵ small avoids overshooting but only at the cost of nodes proceeding slowly to convergence.

When t'' is greater than the threshold T, the node restarts its internal clock $t = 0$ and pulses immediately. Figure 6.1(a) illustrates the evolution of the timer t during a period when the device by itself, i.e., without any neighbors. Without receiving any external pulses, the device clock progresses like a sawtooth waveform with a period equal to 100. But, when the device detects three external pulses indicated by the yellow stars (on the T axis), then it advances its internal clock every time based on the function $f(\cdot)$, thereby shortening its period to pulse faster, as illustrated in Figure 6.1(b).

6.3.2 Network Model

A decentralized CRAHN comprising CR users, communicating across a multi-hop network has been considered. It is assumed that each CR user can discover other CR users in the vicinity, and opportunistically take advantage of the unused portions of the RF spectrum. Each CR user is denoted by X_i, where $1 \le i < n$, being n the total number of CR users. CR employ algorithms and protocols to manage, share and sense the spectrum as described in the literature [1]. However, the design and purpose of the proposed BSynC is to be independent of the choice of these other strategies during the cognitive cycle. Each CR user, X_i has a dedicated receiver to read the beacons on a common control channel (CCC). The CCC is a dedicated channel on which a CR user can (i) discover other CR users, thereby establishing contact; (ii) coordinate the spectrum access; and (iii) identify opportunities in the spectrum.

Every CR user, X_i has an internal clock, t_i, and an estimation of the real time. Each X_i can assume one or more of the following roles during the synchronization phase. They are: *master node*, *ordinary node*, *ordinary node of reference*, and *neighbor node*. These roles are explained as follows.

- A **master node (MN)** is equipped with any device that provides the UTC such as a global position system (GPS). These nodes possess real time information in order to facilitate the convergence of internal clocks across the network. In other words, an MN is another CR user with a GPS. Because deploying many MNs will be energy-inefficient, we assume a very small number of MNs for a network, the details of which are discussed in the later sections. Notation-wise, MNs are denoted by X_i^*. Here, X_i^* refers to MN identified by X_i.
- **Ordinary node (ON)** is a CR user needing to synchronize its internal clock. It synchronizes its clock with another CR user that can be an MN or another CR user that has already synchronized itself, referred to as the ordinary node of reference (RN), described below.
- **Ordinary node of reference (RN)** is a CR user that is not an MN, and whose clock time is used as reference for other ONs.
- **Neighbor node (NN)** is a CR user within the radio coverage of another CR user, considering that those users are using the same CCC. The set of neighbors of a CR user, X_i is denoted by $Neig_i$.

6.4 <u>B</u>io-inspired <u>S</u>yn<u>C</u>hronization Protocol

The synchronization process of the BSynC protocol is initiated by the master nodes and follows two well-defined cyclical phases: (i) **Phase 0**, during which synchronization requests are broadcast over the network, and (ii) **Phase 1**, during which the internal clocks are iteratively adjusted. Spectrum sensing enables the CR users to obtain a list of the "free" channels in an area upon which each user initiates Phase 0. During Phase 0, every ON broadcasts to its NN this list over the CCC. In conjunction with each of its neighbors, the ON defines a common synchronization channel (i.e., a channel free of PU presence for both the sender and the receiver) and they tune to that channel to convey the synchronization messages. Once tuned to the same common synchronization channel, pairs of nodes initiate Phase 1 to determine the difference between real-time clock values of the nodes and the difference between internal timers. Based on the difference between these values, the nodes synchronize their clocks, and when t_i value reaches T, a pulse consisting of broadcasting beacons transmitted, and Phase 0 is reset. This process iterates, after which the nodes will have the same internal clock times.

The proposed BSynC protocol has three key messages: *beacon request*, *beacon response*, and *synchronization*. The beacon request, $b_{req}(L_i)$ contains a list of free channels L_i from a CR user X_i. The beacon response has the value of the common synchronization channel chosen for synchronization between pair of nodes X_i and X_j. The beacon response is denoted by $b_{res}(c_{ij})$ and the common synchronization channel by c_{ij}. Synchronization messages are denoted by $m(X_i, t_i, Ts_i)$ and contains the identifier of the CR user X_i, the value of its internal timer t_i, and its timestamp Ts_i. The first two types of messages are used during Phase 0 of the protocol, while the latter type of message is used during Phase 1. The exchange of messages as beacons is performed over the CCC, whereas the synchronization messages are carried over the common synchronization channel chosen by the neighboring users. Next, the two phases of the protocol are described based on the notation presented in Table 6.1.

6.4.1 Phase 0: Synchronization Request Broadcast

The synchronization request phase is performed by each node in the network. Each MN, X_i^* initializes t_i to 0 and is incremented with the same periodicity of its real-time clock. When t_i becomes equal to T, X_i^* broadcasts the message b_{req} to nodes in $Neig_i$, and resets its internal clock, i.e., sets t_i to 0. Each neighbor X_j receiving $b_{req}(L_i)$ responds with a $b_{res}(c_{ij})$ containing the

Table 6.1   Terminology and definitions of key terms.

Notation	Definition
X_i e X_j	Node identifiers from two arbitrary neighbor nodes X_i and X_j, respectively
X_i^*	Master node
$Neig_i$	A set of neighbors of a given node X_i
L_i	A list of free channels for X_i
c_{ij}	Chosen channel for synchronization between X_i and X_j
$b_{req}(L_i)$	Request beacon having the list of free channels L_i for X_i
$b_{res}(c_{ij})$	Response beacon having the chosen channel c_{ij} by X_i and X_j
t_i	Internal timer for X_i
T	Threshold for synchronization cycle
Ts_i	Timestamp of X_i
$m(X_i, t_i, Ts_i)$	Synchronization message
M	Set of malicious nodes

chosen synchronization channel. After defining a synchronization channel, both X_i^* and X_j tune to c_{ij} and X_i^* sends a $m(X_i, t_i, Ts_i)$. All nodes of $Neig_i$ receiving this message establish X_i^* as its reference node. Then, X_j sets its internal clock during the next phase. After these steps, the node X_j transmits a beacon $b_{req}(L_j)$ for nodes in $Neig_j$. Those nodes in $Neig_j$ establish X_j as their RN.

6.4.2 Phase 1: Setting the Internal Clock

Phase 1 starts when a node X_j receives a message $m(X_i, t_i, Ts_i)$ from a node X_i. This message contains the time stamp from X_i, which is used as reference by X_j. If Ts_j is larger than Ts_i, then X_j decreases the value of its real-clock. Otherwise, if Ts_j is less than Ts_i, X_j increases the value of its real-time clock. X_j also receives X_i's internal clock value, t_i. For this value, X_j also compares the value of t_i with the value of its own internal timer t_j. If t_i and t_j are different, Eq. (6.1) is applied to define t_j, the new value for t_j.

$$t_j' = f^{-1}(f(t_j) + \epsilon), where f(t) = t. \tag{6.2}$$

Figures 6.2 and 6.3 illustrate the BSynC operation. First, MNs (1 and 7) execute Phase 0 by broadcasting the synchronization request beacons to its neighbors $b_{req}(L_1)$ and $b_{req}(L_7)$ containing a list of free channels (2, 3 and 5). Figure 6.3 illustrates an example scenario in which the transmitter node X_i sends a beacon request containing its free channels 2, 4, and 5 ($b_{req}(2, 4, 5)$) to a neighboring node X_j. X_j replies by sending the channel 5, $b_{res}(5)$ that is a free channel and common to both nodes. After setting the common synchronization channel, node 2 and node 3 set node 1 as their reference node,

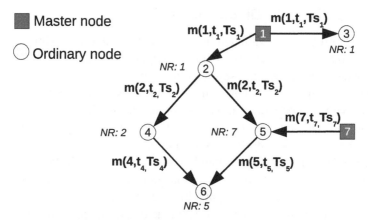

Figure 6.2 Illustration of the Phase 0 in BSynC.

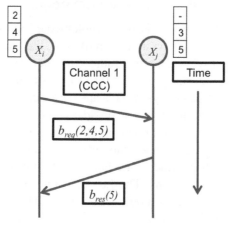

Figure 6.3 Synchronization channel negotiation between a MN (or an RN) and an ON.

and node 5 establishes node 7 as its reference node. The nodes tune to the chosen channel and start Phase 1.

After completing Phase 1, and synchronizing with node 1, node 2 can receive another beacon request from node 5. In this case, to prevent incorrect alterations (moving it forward to slowing it) to its real-time clock value, node 2 sends a query for its reference node (node 1), which responds by sending its time and compares its time with node 5 time. According to this comparison, node 2 increases or decreases its real-time clock. This same process is performed by all other nodes.

6.5 Evaluation and Discussions

The BSynC protocol is evaluated using simulations and its performance was observed and analyzed. The protocol is compared to the TPSN protocol [7], that provides a reference for several other synchronization protocols in the literature, as well as for the CR-Sync [15], the only synchronization protocol for cognitive radio networks in the literature. Further, TPSN presents higher precision and demands less convergence time that other protocols [7]. Also, in CR-Sync paper, the authors present only an analytical model, making the comparison by simulations a demanding task. Differently, TPSN implementation in NS-2 is available to comparison purposes. Hence, we have chosen TPSN to compare results of BSynC, instead of CR-Sync. The details of the simulation setting, metrics evaluation and the results are presented as follows.

6.5.1 Description of the Simulation Environment

Both BSynC and TPSN protocols were implemented in the Network Simulator NS-2.31. For the cognitive radio simulation, the CRAHN module was applied in order to provide cognitive radio characteristics [5]. Each node is equipped with three interfaces IEEE 802.11a (with cognitive radio capabilities in the case of CR nodes), which are (i) a control interface, used to transmit control packets in CCC and broadcast messages to the neighbors, (ii) a receiving interface, used for sensing the channels of spectrum; and (iii) an interface for executing channel handoffs. This approach was applied to define the role of each interface. The number of channels simulated is 10, a fixed value for all simulations. The Pus transmit data by preset channels, being their use defined by a Bernoulli distribution. The presence of a PU on the channel forces SUs to switch.

The number of node in simulation scenarios varied from 20 to 140 CR nodes randomly distributed in a rectangular region of 1000 m × 1000 m. Each scenario has a different simulation time from 400 to 1500 s. The exact time of each simulation is determined by the synchronization convergence of the network. Each plotted point presents a confidence interval of 95%. The values of the other simulation parameters are summarized in Table 6.2.

The following metrics are used to evaluate BSynC protocol performance. *Overhead on the network*: Traffic (in units of control messages) produced due to the synchronization mechanism in proportion to the total data traffic. *Convergence time*: Time a network takes for all nodes get synchronized for at least a period T. This is one of the key performance indicators for synchronization protocols.

Table 6.2 Simulation parameters values.

Radio propagation model	TwoRayGround
Node mobility model	Random waypoint
Type the MAC layer	IEEE 802.11
Length of the receiving line	500
Antenna model	OmniAntenna
Routing protocol	AODV
Number of PUs	2
Constant ϵ in Eq. (6.2)	0.5
Constant T	1 sec.

The network size plays an important role. Not surprisingly, larger networks are expected to converge slower than small networks. Thus, in order to compare the performance of both protocols, we vary the network size from 20 to 140, with an increment of 20 nodes in each scenario. In order to have a proportional network density due to the increasing number of nodes, the grid size is varied accordingly. Another factor evaluated was the influence of the level of node mobility. In the used mobility model (see Table 6.2), each node is positioned randomly in the grid and moves with a random velocity (maximum speed of 10 m/s). To change the levels of node mobility, the delay between the places of destination is varied from 0 to 10 seconds. A shorter pause time means greater mobility.

6.5.2 Key Results

In this subsection, simulation results are presented considering scenarios with *static nodes* and scenarios with *mobile nodes*. Results for each one of the metrics are observed for these two kinds of scenarios, varying the network size in terms of number of nodes, and mobility in terms of the pause time.

Figure 6.4(a) shows the convergence time (total time that gets to synchronize the wide network) under different network sizes. We can observe that BSynC presents a shorter convergence time than TPSN. This can be explained by the fact that TPSN creates a tree structure before performing synchronization, and for a large network, it gets more time to synchronize leaf nodes, i.e., that nodes at the last level of this structure. BSynC seems to be more flexible and facilitates insertion of a new node, since nodes need to synchronize only with neighbor nodes.

Figure 6.5(a) shows results for the overhead generated for both protocols under the variation of the network size (the number of nodes). The overhead generated by BSynC is slightly larger than TPSN for all analyzed network

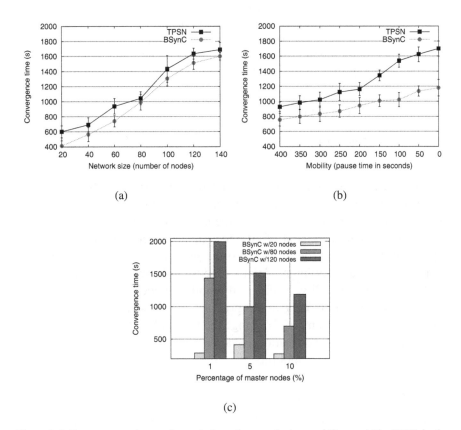

Figure 6.4 Convergence time under variation of *network size*, *mobility*, and *% of MNs* in the CRAHN.

sizes. This behavior is due to the use of a dedicated CCC to exchange messages. Despite the convergence time of BSynC being smaller than TPSN, BSynC looses control data.

While Figures 6.4(a) and 6.5(a) show the results of convergence time and overhead for scenarios with static nodes, Figure 6.4(b) and 6.5(b) compare results considering scenarios with 80 mobile nodes. Figure 6.4(b) compares the protocols under different levels of mobility. The lower is the value of the pause (x-axis of Figure 6.4(b) and 6.5(b)), the greater is the network dynamism. It is observed by Figure 6.4(b) that BSynC presents a considerable improvement in convergence time over TPSN in mobile scenarios. Further,

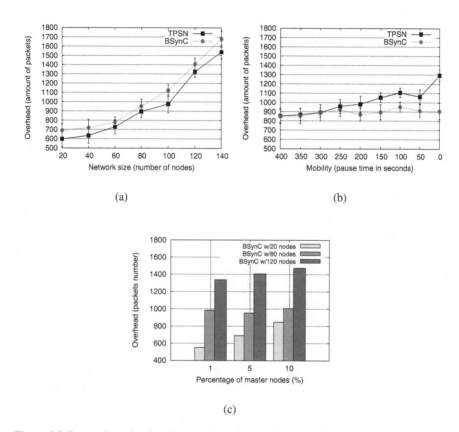

Figure 6.5 Protocol overhead under variation of *network size*, *mobility*, and *% of MNs* in the CRAHN.

the higher network dynamism, the better is the convergence time resulting from BSynC.

Figure 6.5(b) complements Figure 6.4(b) comparing the overhead generated by the protocols under variations in network mobility. The overhead generated by BSynC is less than the overhead incurred by TPSN in scenarios with mobility. It was also observed that contrary what occurs with TPSN, the overhead generated by BSynC remains practically constant with increasing the level of network mobility.

Figures 6.4(c) and 6.5(c) show the convergence time and overhead, respectively, under different percentage of MNs nodes in the CRAHN. Figure 6.4(c) shows convergence time for a network size of 20, 80 and 120 nodes with a 1, 5 and 10% of MNs for each one. The convergence time is smaller while the percentage of master nodes increase, as expected. However, the overhead is affected, as is shown in Figure 6.5(c).

6.5.3 Discussions

Based on the simulation results presented, it can be observed that BSynC yields lower convergence time compared to TPSN in both static and mobile CRAHN scenarios. Furthermore, we have also verified that mobility on the network assists BSynC, more so in dynamic scenarios resulting in better synchronization time and less overhead when compared to TPSN. These results show the flexibility and robustness of the BSynC protocol under changes of network topology. These changes may represent failures in node communications or just temporary discontinuity of the communication. Dynamics on topology may also represent channel changes by secondary users. Comparing the results for the protocol overhead, we observed that BSynC shows a slight increase in overhead in relation to TPSN. This result is justified by the use of a control common channel that can generate losses and delays in synchronization messages due to constraints added by the MAC protocol. However, based on mobile scenarios, the overhead generated by BSynC remains practically constant. This can be explained by the localized operation of the protocol. The exchange of synchronization messages are performed essentially between neighbors and BSynC does not require the formation of a network topology structure, such as tree, thereby reducing the amount of control messages supporting the protocol operation.

6.6 Conclusions and Open Questions

In this chapter, we presented BSynC, a synchronization protocol for CRAHN inspired by the spontaneous firefly synchronization. Unlike the state-of-the-art in this area, BSynC follows no defined network structure and the synchronization between nodes is symmetric. The performance of the novel protocol has been evaluated through systematic simulations, and is compared with TPSN, fundamental to many existing synchronization protocols for wireless ad hoc networks. Results showed that the convergence time of BSynC is smaller than the convergence time of TPSN. Further, our findings suggest that BSynC performs better than TPSN in dynamic contexts, managing efficiently changes in the network topology caused by spectrum handoffs, failures and others. However, different questions are still open. First of all, a better analysis related to the parameter ϵ must be performed in order to define the most adequate value to this parameter considering cognitive radio characteristics. Further, despite of the synchronization model achieves synchronicity among nodes independent of the pulse function used, it is also an open question to determine the better pulse function taking into account the characteristics of cognitive radio networks, their delays and other different factors. In order to improve the proposed approach, it is also planned to eliminate the use of the CCC, and employing another method of exchanging messages to reduce the convergence time and overhead.

References

[1] I. F. Akyildiz, W.-Y. Lee, and K. R. Chowdhury. CRAHNs: Cognitive radio ad hoc networks. *Ad Hoc Netw.*, 7(5):810–836, 2009.

[2] C. Cordeiro and K. Challapali. C-MAC: A cognitive MAC protocol for multi-channel wireless networks. In *Proceedings 2nd International Symposium on New Frontiers in Dynamic Spectrum Access Networks, 2007*, pages 147–157. IEEE, 2007.

[3] F. Dressler and O. B. Akan. A survey on bio-inspired networking. *Comput. Netw.*, 54:881–900, Apr. 2010.

[4] J. Elson and D. Estrin. Time synchronization for wireless sensor networks. In *Proceedings 15th International Parallel and Distributed Processing Symposium*, pages 1965–1970, 2001.

[5] M. Di Felice, K. Chowdhury, and L. Bononi. Modeling and performance evaluation of transmission control protocol over cognitive radio ad hoc networks. In *Proceedings ACM MSWiM*, pages 4–12, 2009.

[6] S. Ganeriwal, R. Kumar, and M. B. Srivastava. Timing-sync protocol for sensor networks. In *Proceedings 1st International Conference on Embedded Networked Sensor Systems*, pages 138–149. ACM, 2003.

[7] S. Ganeriwal, R. Kumar, and M. B. Srivastava. Timing-sync protocol for sensor networks. In *Proceedings ACM SenSys*, pages 138–149, 2003.

[8] S. Ganeriwal, S. Čapkun, C.-C. Han, and M. B. Srivastava. Secure time synchronization service for sensor networks. In *Proceedings ACM WiSe*, 2005.

[9] J. V. Greunen and J. Rabaey. Lightweight time synchronization for sensor networks. In *Proceedings 2nd ACM International Conference on Wireless Sensor Networks and Applications (WSNA'03)*, pages 11–19. ACM, 2003.

[10] S. Haykin. Cognitive radio: Brain-empowered wireless communications. *IEEE Journal on Selected Areas in Communications*, 23(2):201–220, 2005.

[11] Y. R. Kondareddy and P. Agrawal. Synchronized MAC protocol for multi-hop cognitive radio networks. In *Proceedings IEEE International Conference on Communications*, 2008.

[12] R. E. Mirollo and S. H. Strogatz. Synchronization of pulse-coupled biological oscillators. *SIAM J. Appl. Math.*, 50:1645–1662, Nov. 1990.

[13] J. Mitola. Cognitive radio: An integrated agent architecture for software defined radio. Doctor of Technology, Royal Inst. Technol.(KTH), Stockholm, Sweden, pages 271–350, 2000.

[14] M. Morelli and M. Moretti. Robust frequency synchronization for OFDM-based cognitive radio systems. *IEEE Trans. on Wireless Comm.*, 7(12):5346–5355, Dec. 2008.

[15] J. Nieminen, R. Jantti, and Lijun Qian. Time synchronization of cognitive radio networks. In *Proceedings IEEE GLOBECOM*, pages 1–6, Dec. 2009.

[16] P. Pawelczak, R. Venkatesha Prasad, L. Xia, and I. G. M. M. Niemegeers. Cognitive radio emergency networks-requirements and design. In *Proceedings First IEEE International Symposium on New Frontiers in Dynamic Spectrum Access Networks (DySPAN 2005)*, pages 601–606. IEEE, 2005.

[17] Y. Pei, A. T. Hong, and Y. C. Liang. Sensing throughput tradeoff in cognitive radio networks: How frequently should spectrum sensing be carried out? In *Proceedings 18th Annual IEEE International Symposium on Persona, Indoor and Mobile Radio Communication (PIMRC)*, pages 5330–5335, Sept. 2007.

[18] D. Saha, A. Dutta, D. Grunwald, and D. Sicker. Blind synchronization for NC-OFDM: When channels are conventions, not mandates. In *Proceedings IEEE DySPAN*, pages 552–563, May 2011.

[19] M. L. Sichitiu and C. Veerarittiphan. Simple, accurate time synchronization for wireless sensor networks. In *Proceedings IEEE Wireless Communications and Networking (WCNC2003)*, volume 2, pages 1266–1273. IEEE, 2003.

[20] A. Tyrrell and G. Auer. Imposing a reference timing onto firefly synchronization in wireless networks. In *Proceedings Vehicular Technology Conference*, pages 222–226. IEEE, Apr. 2007.

[21] G. Werner-Allen, G. Tewari, A. Patel, M. Welsh, and R. Nagpal. Firefly-inspired sensor network synchronicity with realistic radio effects. In *Proceedings 3rd ACM International Conference on Embedded Networked Sensor Systems (SenSys'05)*, pages 142–153. ACM, New York, 2005.

[22] Y.-C. Wu, Q. Chaudhari, and E. Serpedin. Clock synchronization of wireless sensor networks. *IEEE Signal Processing Magazine*, 28(1):124–138, 2011.

Part III

Security & Tests

Part III

Security & Trust

7

Secure Access Control and Authority Delegation Based on Capability and Context Awareness for Federated IoT

Bayu Anggorojati, Parikshit Narendra Mahalle, Neeli Rashmi Prasad and Ramjee Prasad

Center for TeleInFrastruktur (CTIF), Aalborg University, Denmark
e-mail: ba@es.aau.dk

Abstract

Access control is a critical functionality in Internet of Things (IoT), and it is particularly promising to make access control secure, efficient and generic in a distributed environment. Another important property of an access control system in the IoT is flexibility which can be achieved by access or authority delegation. Delegation mechanisms in access control that have been studied until now have been intended mainly for a system that has no resource constraint, such as a web-based system, which is not very suitable for a highly pervasive system such as IoT. This chapter presents the Capability-based Context Aware Access Control (CCAAC) model including the authority delegation method, along with specification and protocol evaluation intended for federated Machine-to-Machine (M2M)/IoT. By using the identity and capability-based access control approach together with the contextual information and secure federated IoT, this proposed model provides scalability, flexibility, and secure authority delegation for highly distributed system. The protocol evaluation results show that the capability creation and access mechanism of CCAAC is secure against a rigorous man-in-the-middle attack, e.g. eavesdropping and replay attacks, and is able to provide authentication as well.

Fabrice Theoleyre and Ai-Chun Pang (Eds.), Internet of Things and
M2M Communications, 135–160.

Keywords: capability-based access control, delegation, security, IoT.

7.1 Introduction

The Future Internet (FI) will be shaped by the rapid growth of the current internet in terms of penetration, services, capacity and contents along with ultra-small yet powerful mobile communication devices. The vision of internet connectivity by anyone, any time and anywhere, will be augmented by a new dimension, namely connectivity of anything driven by the advancements in the development of smart devices including Wireless Sensor Network (WSN), Radio Frequency Identification (RFID), Near Field Communication (NFC), etc. This is often referred to as IoT and enables different modes of communication, i.e. things-to-person, things-to-things and person-to-things communication along with various intelligent services and applications. A large number of research challenges remain in order to ensure the successful operation of IoT and some of the most significant of these relate to the capabilities of the network to offer security, privacy and trust.

IoT is ubiquitous and heterogeneous in nature. As a result, a large numbers of risks and threats emerge that can threaten the system's security and user's data privacy. Among others, access control is one of the main security issues to be addressed in IoT that is highly pervasive. A novel access control model – CCAAC is proposed to overcome this issue. Furthermore, as IoT is characterized by highly dynamic nodes connectivity and network topologies due to the ever-changing nature of wireless channel, mobility, and factors such as limited power, a dynamic and flexible design of access control system that is suitable for IoT is of the most importance. For this purpose, the proposed CCAAC model also considers the delegation of authority in order to gain access to a certain resource which belongs to different security domain in a Federated IoT network.

The chosen approach for the model with access control based on the capability concept, and in particular the Identity based Capability (ICAP) system [8], is considered in order to cope with the scalability of IoT system since it is well suited for providing authentication and access control in distributed systems. By using the capability concept, the proposed model to some extent also contributes to the authentication mechanism, which will be shown later in Section 7.8. Furthermore, in order to fulfil the flexibility and adaptability in IoT, context awareness and dynamic security as well as privacy policy enforcement is applied in the proposed model. The Context Aware Security Manager (CASM) model originally proposed for a Personal Network (PN)

[18] will be used as a reference point due its suitability in distributed systems, especially in the case of federated IoT. Along with that, flexible access control system by means of authority delegation method based on a dynamic capability propagation is also included in the proposed model [1].

7.2 Related Works on Access Control and Authority Delegation Models

Capability based access control was derived from the capability system concept in the computing world [15] in order to address the shortcomings of Access Control List (ACL) based models. Both capability based access control and ACL are two complementary traditional access control systems derived from an Access Control Matrix (ACM). Basically, an ACM consists of several columns that list the objects or resources to be accessed, and rows that list a set of subjects or users who have privileges to access different resources. Each row in the ACM can also be seen as a capability owned by a subject that needs to be presented whenever it wants to access any object or resource in a system.

A simple example of capability used in the real life is a ticket that allows a passenger to embark on different modes of city transportation, e.g. bus, tram, train, etc. If the ticket contains all the transportation modes in which the passenger can embark, then the capability based access control is used. On contrary, if the system check whether the credential contained in the ticket is listed in the access control list, then the ACL is used. In comparison [3], capability based access control is better than ACL in tackling e.g. the confused deputy problem and other security threats, it is more robust due to its distributed architecture, it supports more levels of granularity, access right delegation and revocation. The Identity based Capability or ICAP was proposed in [8] which solves the problem of capability propagation and revocation in classical capability based access control.

In Context aWare Access Control (CWAC) [13], the surrounding context of subjects and/or objects is considered to provide access but it has scalability issues and no support for delegation and revocation. In Context Aware Role Based Access Control (CARBAC) [14], context is integrated with Role Based Access Control (RBAC) dynamically which is defined as characterization of the surrounding entities for performing appropriate actions. With CARBAC each different role can be associated with certain contexts. For instance, a Professor role in a university can be associated with some contexts, such as

PhD degree, x years of research experience, and y numbers of journal publications in the related field. However, improper association of context and roles results in inefficiency in scalability and time and context dependency results in complicated delegation.

A Personal Trusted Device (PTD) [2] is used to provide access to locked resources and places, which consists of access controllers, the PTD and an administration point, while embedded security of the PTD is not considered. In [16], an access control model is designed as standalone components each managing its context and thus limiting the reuse and context sharing. In [5, 9] solutions are proposed for decoupling the applications from the context resources which require permanent access to the server. However, in a Wireless Personal Area Network (WPAN) as one scenario considered in PN, access to fixed infrastructure cannot be guaranteed due to constraints like frequently changing connectivity, resource unavailability and battery power. In [17], an identity establishment scheme for IoT as well as authentication mechanism based on Elliptic Curve Cryptography (ECC) and capability was proposed.

Research on delegation of authority using capability has been investigated in [11, 10]. Hasebe et al. [11] addressed the issue of role and/or permission delegation based on a RBAC model in a cross-domain environment using capabilities. The central idea behind their proposed mechanisms was the mapping of capabilities into roles and permissions in each domain. Hasebe et al. extends the approach presented in [11] by adding the delegation of task to be performed in the model for workflow systems in [10].

Research on dynamic authorization delegation in federated environment between entities or machine-to-machine delegation has been reported by Gomi et al. in [7]. Authors of the paper investigated chain of delegation in multiple entities and how to provide secure delegation framework. To better illustrate the delegation chain in multiple entities being investigated in [7], we can think of a user who is authorized in his company's web server and wanting to access some information from other company's server, i.e. destination server. In this scenario, his request has to go through his company's server and other intermediate servers in between before reaching the destination server. Consequently, the user has to delegate his authority over the chains of intermediate servers in a relay-like fashion. The delegation framework proposed in [7] also introduced Delegation Authority and Authentication Authority entities to enable such authorization delegation. Another related work in this topic has also been reported by the same author, Gomi, in [6]. Unlike his previous work in [7], in [6] the author focuses on the user-to-user delegation and does not consider multiple entities delegation. The main contribution was a delegation framework in federated environment using an access token, regardless of the access control model being used. In this context, an access token is an opaque string representing a delegator's authorization to delegate their privilege to a delegatee whom they specify, which does not contain any information about delegator's credentials. The paper also explains the mechanism of issuing a token, asserting it into an authorization document, and service provisioning based on the delegate token.

The existing delegation methods are designed mainly to serve web-based services which involve a large IT infrastructure. These kinds of delegation models are not practically visible for an IoT system, which has a lot of constraints, e.g. power, memory, etc. It is important to mention that the delegation of authority by means of capability propagation is part of CCAAC overall design model. Therefore, delegation method in CCAAC is not an extension of any existing access control model, e.g. RBAC, as presented in most of the previous works.

The federated IoT networking has not been much discussed in the literature. There is a significant amount of work done in the context of a federated device-to-device or machine-to-machine communication known as PN federation, which includes the networking, management, security framework and identity management in [18]. Federated IoT environments based on the PN concept [18] will be further elaborated in Section 7.4.2.

7.3 Contributions to the Existing Works

According to the review of the related works on access control and authority delegation, some challenges are still left open, especially in the context of IoT. The biggest challenge on the access control for IoT are the scalability and flexibility, not to mention the secure access control. As we understand already that the IoT is highly pervasive and distributed in nature, thus scalability aspect of access control system is one important characteristic to have. While access control system is usually centralized or "distributed" in a particular entity, e.g. server, our approach is to distribute the complexity to the accessing entity by using capability. The flexibility challenge in access control here refers to two things: the flexibility in the process of access decision making by considering the contextual information and the flexibility in terms of authority delegation. Based on the review of challenges related to such requirements in Section 7.2, the proposed access control scheme will tackle both aspects at the same time.

Figure 7.1   System architecture for supporting the CCAAC.

7.4 System Architecture

7.4.1 System Architecture to Support the CCAAC Model

The system architecture for supporting the CCAAC model is depicted in Figure 7.1. It is important to note that the Personal Network (PN) [18] has been referred in this work due to its advance networking concept for device-to-device communication that opens a path as one candidate of network implementation in IoT. Correspondingly, the security framework brought up within the PN would be a good starting point in designing a security framework, considering their similarities in characteristics and requirements.

Security Decision Point (SDP) in Figure 7.1 is the most important module in the system architecture to receive the capability creation or access request and make the final decision of the request. *Policies* serves as Policies Repository that consists of a collection of various policies for accessing available resources or objects. *Profiles* serves as Profile Repository, which essentially consists of subject as well as object profiles. Both of these components will be referred to as Policies Repository and Profiles Repository later on in Section 7.5. Finally, *Access List* plays an important role in supporting the capability-based authority delegation of access control by controlling the capability propagation as well as revocation through maintenance of a propagation tree as proposed in [8].

In principle, a new propagation tree is created at the *Access List* whenever a new capability is created, with the original capability owner becomes the root of the tree. When the capability owner delegates his authority by propagating his capability to someone else, then the propagation tree is updated by adding a new node. The one who propagated the capability can revoke the authority delegated to someone else before by requesting a revocation message to *Access List*. If the revocation message is valid, the corresponding node will be removed from the propagation tree.

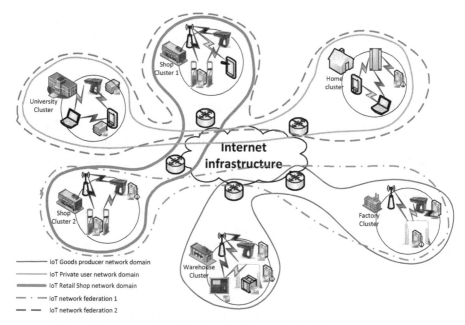

Figure 7.2 An example of Federated IoT network with delegation scenario.

7.4.2 Federated-IoT

Identity Federation is a known term within the web security world and refers to management of a web user's identity across different security domains. The main reason of enabling federation in the web environment is that the work flow of the system often requires a user that is authenticated in one domain to be authenticated in other domains as well. However, a concrete definition of the Federated IoT which is conceptually different as compared to the web security world, needs to be determined before addressing the issue of authority delegation in such environment.

First of all, the identity in the web-based system refers to a person's identity while in IoT, identity refers to a device or "thing". Therefore, the interaction of identities in IoT is in the form of device-to-device communication.

An example of federated IoT network in our context is depicted in Figure 7.2. Three IoT network domains are considered, i.e. private user, retail shop, and goods producer network, and two IoT-Federated networks are considered. An IoT network domain consists of one or more IoT cluster and the inter-cluster communication can be done either through the Internet infrastructure as shown in Figure 7.2 or through a wireless ad-hoc connection. Device-to-device communication within a cluster, i.e. intra-cluster communication, can be carried out by using different wireless access technology, e.g. RFID, ZigBee, bluetooth, wifi, etc.

7.4.2.1 Motivating Scenario for Access Delegation

A good scenario example of how an authority delegation is needed in the Federated IoT networks can be shown as an inter-working of an intelligent-shopping service offered by a retail shop and a smart-fridge device that belongs to *private user* domain. The retail shop offers an intelligent service to its customer such that it is able to suggest a list of foods items based on their availability or expiration dates in the smart-fridge. In order for the retail shop to be able to come up with list of suggestions, it needs a privilege to access the smart-fridge which belongs to different network domain. To solve this issue, a valid privilege will be delegated to a device belongs to *retail shop* network. Before access delegation can be performed, it is assumed that the IoT network federation has been created, hence network authentication and trust relationship between two network domains have been established.

Another scenario of the authority delegation is when a device owned by a *private user* is trying to access the shop cluster 1 which belongs to *retail shop* network domain. In this scenario, suppose that a user that has a "smart fridge" at home is currently at the retail shop 1. He is checking the food at home that needs to be resupplied, at the same time checking those foods that may be available in the shop by accessing any wireless network device in retail shop 1. Since he does not belong to *retail shop* network domain, he needs a privilege in order to read information required by him. To solve such problem, a device that belongs to shop 1 cluster can delegate its privilege to private user's device. Before access delegation can be performed, it is assumed that the IoT network federation has been created, hence network authentication and trust relationship between two network domains have been established. The proposed delegation mechanism and the infrastructure to support the authority delegation will be presented in Section 7.7.

In such authority delegation scenario, some privacy problems could arise. For example, the *private user* do not want to reveal all food or drink items in his smart-fridge to the *retail shop*. With the authority delegation, the *private user* can delegate some of his access rights to *retail shop*. It should be noted that delegating authority does not mean to give full authority to other party as if the other party has the same role as the one delegating it. Furthermore, the delegated authority can be revoked at any time.

Table 7.1 Notations used in the CCAAC definition.

$\mathcal{S}$	Representing identifier of *Subject* that requests an access.
$\mathcal{O}$	Name of object or resource to be accessed.
$\mathcal{AR}$	Type of access right, e.g. read, write, execute.
$\mathcal{C}$	Context information.
Rnd	Random number generated from a one-way hash function to prevent forgery.
$_{ext}CAP$	External capability that is held by the *Subject*.
$_{in}CAP$	Internal capability that is kept in the *Object*, server, or whatever physical entity that is be accessed.
VID	Virtual Identity.
$\mathcal{ID}$	Unique identifier of the *Subject*.
$\mathcal{P}$	A set of *Subject*'s profiles.
C	A set of contextual information that are used to define the VID and *Policy*.

7.5 Proposed CCAAC Model

The proposed CCAAC model also includes properties of ICAP and context aware security, especially the concept of CASM and Virtual Identity (VID) [18]. This fills the gaps in current solutions as emphasized in Section 7.1.

7.5.1 Proposed Capability Structure in CCAAC

In the remainder of this chapter, the notations listed in Table 7.1 will be used.

Firstly, in order to support the context awareness in ICAP, an additional field is added in the $_{ext}CAP$ for the Subject S_i that contains the context information, $\mathcal{C}$, related to the capability. By including this, the external capability structure in CCAAC is defined as follows:

$$_{ext}CAP_i = \{\mathcal{O}, \mathcal{AR}, \mathcal{C}, Rnd_i\} \qquad (7.1)$$

where

$$Rnd_i = f(\mathcal{S}, \mathcal{O}, \mathcal{AR}, Rnd_0) \qquad (7.2)$$

$$Rnd_0 = f(\mathcal{O}, \mathcal{AR}) \qquad (7.3)$$

The internal capability ($_{in}CAP$) that creates a pair with the $_{ext}CAP$ which is stored in the object itself or an entity that has higher "authority" over the object (e.g. in hierarchical type of network), is defined as follows:

$$_{in}CAP = \{\mathcal{O}, Rnd_0\} \qquad (7.4)$$

where Rnd_0 is defined exactly as in Equation (7.3).

7.5.2 Basic Definitions

The important definitions used in CCAAC are presented in this section in which PN is the targeted platform.

7.5.2.1 Contexts

The *Contexts* $\mathcal{C}$, that are used to define the VID and *Policy* is essentially a set of contexts ($\mathcal{C}_{Set}$) with different types ($\mathcal{C}_{Type}$). The type of context can be a concrete property such as time or location, but also security related context such as authentication and trust level. In order to apply the context in the access control decision, each of the context types has to be evaluated with a certain constraint ($\mathcal{C}_{Const}$).

The overall context definition in CCAAC can be expressed with the following notation:

$$\mathcal{C}_{Type} \in \{authLevel, trustLevel, time, location, \cdots\} \qquad (7.5)$$

$$\mathcal{C}_{Set} = \{\mathcal{C}_{Type(1)}, \mathcal{C}_{Type(2)}, \cdots, \mathcal{C}_{Type(n)}\} \qquad (7.6)$$

$$\mathcal{C}_{Const} := \langle \mathcal{C}_{Type} \rangle \langle OP \rangle \langle VALUE \rangle \qquad (7.7)$$

where OP is a logical operator, i.e. $OP \in \{>, \geq, <, \leq, =, \neq\}$ and $VALUE$ is a specific value of $\mathcal{C}_{Type}$. Finally, we defined $\mathcal{C}$ as a set of context constraint $\mathcal{C}_{Const}$ as follows:

$$\mathcal{C} = \{\mathcal{C}_{Const(1)}, \mathcal{C}_{Const(2)}, \cdots, \mathcal{C}_{Const(n)}\} \qquad (7.8)$$

7.5.2.2 Policy

A policy is associated with certain VID(s) that describes VID(s) preferences upon allowing other entities to access them. Please note that the entity or subject requesting access is described by its profile, e.g. subject's attributes, in the policy. A policy in the proposed CCAAC model holds an important role in access control decision as well as any process involving capability creation and delegation. Hence, it can simply be defined as a set of rules with parameters related to the user as follows:

$$Policy \in \{\mathcal{P}, \mathcal{C}, \mathcal{AR}\} \qquad (7.9)$$

It is important to note that since a *Policy* is linked with a VID, and a VID itself is already linked with an *Object*, therefore the *Object* $\mathcal{O}$ notation is not included in the *Policy* statement. Furthermore, unlike the definition of the VID, the $\mathcal{P}$ and $\mathcal{C}$ that are included in the *Policy* statement are related to the *Subject* $\mathcal{S}$ who tries to access the *Object* $\mathcal{O}$.

7.5.2.3 VID

An entity, i.e. subject or object, may have more than one identity, namely one main identity and numbers as other alias identities; this is referred as VID. A VID consists of a user identifier and a set of disclosure policies where the same disclosure policy can apply to different VIDs. Therefore, the relationship between VIDs and disclosure policies is a many-to-many, which can be implemented using a pointer or hash map.

A VID is also attached with a particular context as well as profile information of the corresponding user [18] and it can be assumed that the profile information can be pre-defined as a set of default profiles or customized to a specific VID. The profile is assumed to have an one-to-one relationship with the VID and, for the context, to have a many-to-many relationship with the VID.

Based on these relationships and assumptions, the VID is defined as follows:

$$VID \in \{\mathcal{ID}, \mathcal{P}, \mathcal{C}, Policies\} \qquad (7.10)$$

The *Profile* $\mathcal{P}$ in VID may consist of *Objects'* $\mathcal{O}$ attributes and personal information. $\mathcal{C}$ refers to *Contexts* that can be a security context, such as trust level, authentication level as well as other common contexts such as time and location (see Subsection 7.5.2.1). The $\mathcal{ID}$ is a unique identifier that can be acquired through cryptographic operations. The *Policies* is a set of *Policy* that was explained in 7.5.2.2.

From the practical point of view, the "attributes" attached with the VID, *Policies*, $\mathcal{P}$, and $\mathcal{C}$ do not necessarily need to be physically contained within the VID itself. The VID could for instance contain just the pointers to the relevant *Policies* or *Profile*, and be implemented using overlay Distributed Hash Table (DHT) approach that is commonly used for sharing the information in the Peer-to-Peer (P2P) or ad-hoc networks in a distributed manner. Nevertheless, the practical implementation of the VID as well as the networking and communication protocols used among the communicating entities are outside the scope of this chapter.

7.5.2.4 Other Definitions

Other definitions that are used in the formal specification of CCAAC are as follows:

$$\mathcal{P} = \{Profile_1, Profile_2, \cdots, Profile_n\} \qquad (7.11)$$
$$Policies = \{Policy_1, Policy_2, \cdots, Policy_n\} \qquad (7.12)$$

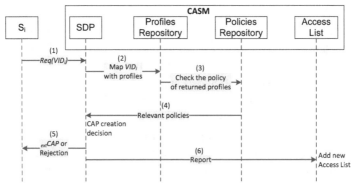

Figure 7.3 Capability creation protocol in the proposed access control mechanism.

$$\mathcal{AR} \in \{Read, Write, NULL\} \qquad (7.13)$$

Note that $\mathcal{P}$ is a set of entity's *Profile* contained in the *VID*, and is stored in the Profiles Repository. The same goes for *Policies*, it is a set of any entity's *Policy* and is stored in the Policies Repository. The usage of both Profiles and Policies Repository will be shown when the detail mechanisms of CCAAC will be explained in the next sub section. As for the *AR*, it can either be $\{Read\}$, $\{Write\}$, $\{Read, Write\}$, or $\{NULL\}$. If $AR = \{NULL\}$, the permission to access a particular object is not allowed.

7.6 Specification of CCAAC Mechanisms

In the following, the formal specification for two of the four main processes in the CCAAC secure access control mechanism is given in details, i.e. the capability creation and access. The specification assumes that CASM is used and the message exchanges among the internal modules are as depicted in Figure 7.1. In the proposed CCAAC model, the SDP is not only acting as a central point for security decisions as was the case in CASM, but is also responsible of performing other functionalities as explained in the following.

7.6.1 Creation

The capability creation protocol in the proposed access control mechanism is presented in Figure 7.3 and can be explained as follows:

1. The *Subject* (**S**) sends a capability creation request of a certain *Object* (**O**) along with its own VID (VID_S) and the VID of the **O** (VID_O) to

be accessed are sent by **S** to the SDP. As discussed earlier, **S** accesses **O** according the VID_O that can be "seen" by the **O**. Whereas the VID_S will be used later by the **O** to obtain $\mathcal{P}$ and $\mathcal{C}$ related to **S** later on.

2. Upon receiving this request, the SDP asks the Profiles Repository to map the profiles of **S** given its VID_S. This process will return the Profile $\mathcal{P}$ of **S**.

3. The returned Profile $\mathcal{P}$, together with $\mathcal{C}$, are sent to the Policies Repository, to check the relevant *Policies* of the corresponding *Object* **O** (based on its VID_O).

4. The Policies repository gets all relevant *Policies* from the given $\mathcal{P}$ and $\mathcal{C}$ of the **O** of interests represented by its VID_O, and then gives them to the SDP.

5. The SDP combines the received policies with a policy combining algorithm and comes up with a decision whether to create a new capability (CAP) for **S** or not. In case of positive decision, the SDP creates internal capability of the object ($_{in}CAP$) and stores it in the CASM, as well as creating $_{ext}CAP$ for the **S** and then sends it away. In case of negative decision, rejection message will be sent to **S** instead.

6. The SDP sends a report regarding the CAP creation of an *Object* **O** for the *Subject* **S** to the Access Control Servers (ACS) which will be followed by the creation of a new propagation tree.

The specification of the capability creation in CCAAC can be expressed by two pseudo-codes. The first pseudo-code presented in Algorithm 1 requires a specific *ObjType* as the argument and returns either an $_{ext}CAP$ to the subject or an error message if certain conditions are not fulfilled. The other pseudo-code, which is presented in Algorithm 2, does not require a specific *ObjType* as its argument and returns either a set of external capabilities or an error message to the subject.

7.6.2 Access Provision

The capability access protocol in the proposed access control mechanism is presented in Figure 7.4. The access mechanism based on CCAAC is relying fully on the processing of the $_{ext}CAP$, i.e. $_{ext}CAP$ validation and evaluation of *Contexts* $\mathcal{C}$ within the $_{ext}CAP$, thus it results in an efficient access control mechanism. Furthermore, the authentication can also be achieved along with the access control as we will explain later on in the next section. The overall proposed access mechanism in CCAAC can be further explained as follows:

Algorithm 1 Capability creation for a particular object type

1:   **procedure** CAPCREATION(VID_S, VID_O)
2:       $\mathcal{P} \leftarrow getProfiles(VID_S)$
3:       $\mathcal{C} \leftarrow getContexts(VID_S)$
4:       $Policies \leftarrow checkPolicies(\mathcal{P}, \mathcal{C}, VID_O)$
5:       $\mathcal{AR} \leftarrow combinePolicies(Policies)$
6:       **if** $\mathcal{AR} \neq NULL$ **then**
7:          **if** $_{in}CAP == 0$ **then**
8:            $_{in}CAP \leftarrow createIntCAP(VID_O)$
9:          **end if**
10:        $_{ext}CAP \leftarrow createExtCAP(VID_S, \mathcal{AR}, VID_O)$
11:       $sendExtCAP$ *to* **S**
12:      **else**
13:        $Send\_Err$ *to* **S**
14:      **end if**
15: **end procedure**

Algorithm 2 Capability creation in the general case

---

  procedure CAPCREATION(VID_S)
     $MapAR < \mathcal{O}, \mathcal{AR} >= [\,]$
     $\mathcal{P} \leftarrow getProfiles(VID_S)$
     $\mathcal{C} \leftarrow getContexts(VID_S)$
     $Policies \leftarrow checkPolicies(\mathcal{P}, \mathcal{C})$
     $MapAR < \mathcal{O}, \mathcal{AR} > \leftarrow combinePolicies(Policies)$
     if $MapAR \neq NULL$ **then**
       for all $\mathcal{O}$ *in* $MapAR$ **do**
          if $\mathcal{AR} \neq NULL$ **then**
            if $_{in}CAP == 0$ **then**
              $_{in}CAP \leftarrow createIntCAP(\mathcal{O})$
            end if
            $_{ext}CAP \leftarrow createExtCAP(VID_S, \mathcal{AR}, \mathcal{O})$
            $SendExtCAP$ *to* **S**
          else
            $SendErr$ *to* **S**
          end if
       end for
     end if
  end procedure

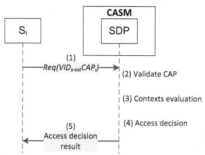

Figure 7.4 Capability access protocol in the proposed access control mechanism.

1. The *Subject* **S** presents its $_{ext}CAP_S$ and VID_S to the SDP upon access request.
2. SDP checks the validity of the $_{ext}CAP_S$ by running a one-way hash function $f(S, \mathcal{O}, \mathcal{AR}, Rnd_0)$ and the compare its result with Rnd_i within the $_{ext}CAP_S$.
3. In case of a valid $_{ext}CAP_S$ being received, SDP will then evaluate a series of $\mathcal{C}_{Const}$ which are contained within the $\mathcal{C}$.
4. The evaluation will return *true* if all the conditions in all of the $\mathcal{C}_{Const}$ are met, otherwise it will return false.
5. The access request response will finally sent back to **S**.

The $_{ext}CAP$ validation (step (2) in Figure 7.4) is done by comparing the received Rnd_i within the $_{ext}CAP$ with the calculated Rnd_i by means of a one-way hash function $f(S, \mathcal{O}, \mathcal{AR}, Rnd_0)$. The most important components in the hash function of Rnd_i to be highlighted are the S and Rnd_0. The S can be obtained from the S within the submitted VID_S and the Rnd_0 is obtained from the $_{in}CAP$ that is stored within the **O** itself. A secure access control against eavesdropping or replay attacks can be ensured by inserting a nonce within the Rnd_0 as will be shown in Section 7.8.

After the $_{ext}CAP$ validation is successfully done, the actual contextual information will be evaluated against *Contexts* $\mathcal{C}$ within $_{ext}CAP$ (step (3) in Figure 7.4). It is assumed that the **O**, i.e. device, is able to obtained the actual contextual information from the environment, e.g. sensors or time of access, or by means of messaging mechanisms.

The specification of the access mechanism in CCAAC can be expressed as in the pseudo-code presented in Algorithm 3.

Algorithm 3 Access control with external capability

---

  procedure ACCESS(VID_S, $_{ext}CAP$)
    $AR' = [\,]$
    $Rnd'_i = f(\mathcal{S}, \mathcal{O}, \mathcal{AR}, Rnd_0)$
    if $(Rnd'_i\,! = Rnd_i)$ **then**
      $Send\,Err\,to\,\textbf{S}$
    else
      $CtxEval = true$
      for all $\mathcal{C}_{Const}$ *in* $\mathcal{C}$ **do**
        $CtxEval = CtxEval \cap \mathcal{C}_{Const}$
        if $CtxEval$ **then**
          $AR' = AR$
        else
          $AR' = NULL$
        end if
      end for
    end if
  end procedure

7.7 Secure CCAAC Based Delegation Framework

7.7.1 High Level Delegation Model

To support a federation network in IoT, it is necessary to have an established trust relationships prior to the authority delegation of all the entities involved in this process. The existing solutions in [18] allow all Federation members (in the context of PN) to mutually authenticate to each other, thus establishing trust relationships, by means of key management and security associations mechanisms. By using the similar approach in the context of federated IoT, it is assumed that the trust relationships have been established in all entities joining the federation when it is formed and any new cluster join that federation. Based on this assumption, the proposed high level delegation model can be illustrated as in Figure 7.5. In this figure, *Delegator* is an entity that delegates some or all of its authority to another entity, while *delegatee* is an entity that receives an authority delegation from the *delegator*. According to the first motivating scenario stated in Section 7.4.2, the user's private device that has the authority to access the information from the smart-fridge is the *delegator* that delegates its authority partly to the *delegatee* which is the device belongs to retail shop network domain.

    Please note that in CCAAC notation, terms such as *Subject* and *Object* is defined. Any *Subject* can be either *Delegator* or *Delegatee*, but *Delegator*

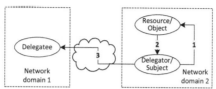

Figure 7.5 High level delegation model on Federated-IoT.

will be referred to as *Subject* (**S**) for the sake of the protocol explanation. The resource to be accessed by *Subject* is referred as *Object* (**O**).

A trust relationship is assumed to have been established among all the entities when the federated network is created as discussed extensively in [18]. Therefore, **S** sends a delegation request to **O**. The delegation request is signed with its public key that contains a federated IoT certificate, which is valid in this particular federation, upon requesting its authority delegation towards **D** (step 1 in Figure 7.5). Upon receiving the delegation request from **S**, **O** would verify the signature and then evaluate the delegation request based on some available policies which will be further explained in the next subsection. In case of a positive delegation request evaluation result, a delegation request response in a form of external capability ($_{ext}CAP_D$) with **D**'s identity would be sent to **S**, otherwise an error message would be sent instead (step 2 in Figure 7.5). Finally, the **S** would send the $_{ext}CAP_D$ encrypted with a public key that is known by **D** as a result of trust relationship when federated network between two domains was established (step 3 in Figure 7.5).

7.7.2 Delegation Mechanism Based on CCAAC

The complete process of the delegation mechanism along with the delegation request evaluation in **O** is depicted in Figure 7.6.

It is important to mention that Figure 7.6 is the micro-level view of Figure 7.5 where **S** and **O** belong to Network Domain 2, and **D** belongs to Network Domain 1. The delegation mechanism depicted in Figure 7.6 is presented as follows:

1. **Sending authority delegation request:** Authority delegation request is being sent by **S** to the SDP within **O**. The request message is signed with **S**'s public that contains a Federated IoT certificate so that **O** is able to make sure the message is indeed sent by **S** and the integrity is maintained. Please note that the type of public key and the specific encryption algorithm being used are not within the scope of this work.

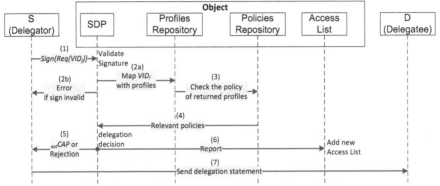

Figure 7.6 Capability propagation protocol for authority delegation in our proposed access control.

2. **Mapping the VID_D to Profile:** The SDP checks the message's signature. If the signature is valid, the SDP then asks VID-Profiles mapping box to map the profiles of **D** given VID_D. It will return the profile of **D**.

3. **Check the relevant policies:** The returned Profiles $\mathcal{P}$, together with the $\mathcal{C}$ and VID_O, are then sent to the Policies Repository, to check the disclosure policies of the corresponding Object (based on its VID).

4. **Return the relevant policies:** The Policies Repository gets all the relevant policies from the given $\mathcal{P}$ and $\mathcal{C}$ of the object or resource of interest represented by its VID_O, and then gives them to the SDP.

5. **Delegation decision:** The SDP combines the received policies with a policy-combining algorithm and comes up with a decision whether to approve the authority delegation by creating a new capability (CAP) for **D** or not. In the case of a positive decision, the SDP creates a delegation statement in the form of $_{ext}CAP$ for the **D** and then sends it to **S**. The difference between the newly created $_{ext}CAP$ and the one that is owned by the **S**, lies in the **D**'s identifier $\mathcal{D}$ within the Rnd_i component. In the case of a negative decision, a rejection message will be sent to **S** instead.

6. **Update propagation tree:** In parallel to sending the delegation decision, the SDP sends a report regarding the CAP creation of an object for Subject i, S_i, to the ACS which will be followed by the creation of a new propagation tree.

7. **Sending authority delegation statement:** Finally, **S** sends the authority delegation statement in the form of $_{ext}CAP$ particularly for **D**. Moreover, in order to maintain the confidentiality and integrity of the $_{ext}CAP$, it can be signed with a shared secret key between **S** and **D**,

based on an assumption that both domains have established a trust relationship by authenticating each other through a certain key pair.

Steps 2 through 5 is expressed in the pseudo-code in Algorithm 4.

Algorithm 4 Capability delegation decision

---

procedure DELEGATECAP(VID_D, VID_O)
    $\mathcal{P} \leftarrow getProfiles(VID_D)$
    $\mathcal{C} \leftarrow getContexts(VID_D)$
    $Policies \leftarrow checkPolicies(\mathcal{P}, \mathcal{C}, VID_O)$
    $decision \leftarrow combinePolicies(Policies)$
    if $decision \neq NULL$ **then**
        if $IntCAP = 0$ **then**
            $_{in}\mathcal{CAP} \leftarrow createIntCAP(VID_O)$
        end if
        $_{ext}\mathcal{CAP} \leftarrow createExtCAP(VID_D, \mathcal{AR}, VID_O)$
    end if
end procedure

It is assumed that **S** as the *delegator* knows the identity of the **O** and **D**. This is possible when the *delegator* **S**, subscribes to a service, in which a device in the service provider's domain needs to be given an authority delegation, i.e. as *delegatee* **D**, in order to access a device or resource within the *delegator*'s network domain, i.e. **O**. With this assumption, **S** needs to submit a delegation request by stating the identities of **D** and **O** in the form of VID_D and VID_O, respectively. Once submitted, *delegatee*'s Profile ($\mathcal{P}$) and Context ($\mathcal{C}$) can be obtained from VID_D as they are attached to it (see Equation 7.10). Afterwards, all relevant policies related to VID_O that contain *delegatee*'s $\mathcal{P}$ and $\mathcal{C}$ are gathered from the Policies Repository to be further evaluated by a certain Policies Combining Algorithm to obtain a delegation decision.

7.8 Evaluation and Discussion

The evaluation will focus on secure capability creation and access mechanisms as the most important processes in the access control, especially when capability is involved. In order to secure the access control mechanism, simple mechanisms of generating nonce in both sides and a secret key cryptography to encrypt the message are introduced. The Automated Validation of Internet Security Protocols and Applications (AVISPA) tool [12] which is based on the Dolev–Yao [4] model is used for model verification purposes as well as for evaluating the secrecy and authentication between the subject, i.e. the one that requests access, and the object, i.e. the one that is being accessed.

7.8.1 Evaluation Procedure

In order to carry out the evaluation using AVISPA, the following assumptions are made:

- Both *Subject* and *Object* have already obtained shared key(s), i.e. either symmetric or asymmetric, through key generation and sharing mechanisms prior to the capability creation process.
- To prevent a replay attack, a nonce is incorporated within the Rnd_0 of $_{in}CAP$ and $_{ext}CAP$.
- A new key, which is a result of a one-way hash function using nonce generated by both parties, is presented by **S** that wants to access **O**.
- The access mechanism using external capability is used immediately after the capability creation in order to have an integrated evaluation of both mechanisms.

  The complete protocol evaluation is presented in the following model:

  $$\mathbf{S} \rightarrow \mathbf{O}: \quad \{S.Req.N_s\}\_K_{so}$$
  $$\mathbf{S} \leftarrow \mathbf{O}: \quad \{S.\mathcal{O}_{xo}.\mathcal{AR}_s.\mathcal{C}_s.N_o.Rnd_i\}\_K_{so} \qquad \text{where}$$
  $$\mathbf{S} \rightarrow \mathbf{O}: \quad \{S.\mathcal{O}_{xo}.\mathcal{AR}_s.\mathcal{C}_s.Rnd_i\}\_K_N$$

- S: Subject identifier.
- $\mathcal{O}$: Object identifier.
- $\{ \ \}\_$: A symbol of encryption.
- Req: A capability creation request message.
- N_s: A nonce generated by **S**.
- N_o: A nonce generated by **O**.
- K_{so}: A shared secret key shared by the **S** and **O** prior to the capability creation process to encrypt the whole message.
- K_N: A new shared secret key generated by the **S** as a result of a one-way hash function, $f(N_s.N_o)$.
- Rnd_i: A result of a one-way hash function $f(S.\mathcal{O}_{xb}.\mathcal{AR}_s.\mathcal{C}_s.Rnd_0)$.
- Rnd_0: A result of a one-way hash function $f(\mathcal{O}_{xb}.\mathcal{AR}_s.N_o)$.

  Besides the protocol that involves **S** and **O**, an intruder, **I**, based on Dolev–Yao intruder model has been introduced in the evaluation. The intruder **I** is assumed to have the knowledge of the following:

- S: the Subject's identifier.
- K_{si}: the shared secret key between **S** and **I**.

- K_{io}: the shared secret key between **I** and **O**.
- all the hash functions being used in the protocol, i.e. $f(N_s.N_o)$, $f(\mathcal{S}.\mathcal{O}_{xo}.\mathcal{AR}_s.\mathcal{C}_s.Rnd_0)$, and $f(\mathcal{O}_{xo}.\mathcal{AR}_s.N_o)$.

The goal of the evaluation is to verify the secrecy of the new generated key presented by **S**, e.g. K_N, in order to access **O**, and the N_o generated by **O** in order to create Rnd_0. Furthermore, N_o is also used to authenticate **S** over **O**. These two important aspects in the evaluation will be discussed further in the coming subsections.

7.8.2 Evaluation and Discussion of the Secrecy

The secrecy of N_o is an important component to prevent replay attacks on the $_{ext}CAP$ that is submitted by **S**. The evaluation result using the AVISPA tool shows that the secrecy of N_o is kept. It is important to note that the Dolev–Yao intruder model used in the evaluation is assumed to have knowledge of Subject's identifier $\mathcal{S}$, the hash functions being used, and the shared keys, K_{si} and K_{io}. The intruder cannot decrypt the message being sent by either side even if he can capture it since he has no knowledge of K_{so}. In other words, the knowledge of K_{si} and K_{io} that might be used by the intruder to fool both sides will be useless in this case. As a result, both nonce, i.e. N_s and N_o, remain secret. Secondly, the secrecy of K_N which is a result of simple one-way hash function $f(N_s.N_o)$ as presented in the evaluation model is another important fact to be evaluated. Similar with the N_o, the evaluation result by AVISPA tool also shows that K_N remains secret. As explained previously, the secrecy of K_N is a result of the secrecy of both N_s and N_o due to inability of the intruder to decrypt the message without having the knowledge of K_{so}. Based on this evaluation result, i.e. the fact that K_N remains secret, it could be concluded that the $_{ext}CAP$ is safe from eavesdropping and replay attacks.

7.8.3 Evaluation and Discussion of the Authentication

The authentication of **S** against **O** can be achieved when **O** receives a valid Rnd_i. In particular, the received N_o within the Rnd_i, which is supposed to be a secret nonce generated by **O** and known only by **S**, has to be valid and kept secret. The evaluation result by AVISPA tool shows that authentication is achieved as well. As stated previously in the secrecy analysis, the fact that the secrecy of N_o that can be ensured in the protocol results in the validity of Rnd_i within the $_{ext}CAP$, thus **S** can successfully authenticated by **O**. Furthermore, this result also shows that the proposed CCAAC model is not only able to provide access control but also authentication at the same time. Hence, the main objective of providing security mechanisms, i.e. authentication and access control, in order to prevent security threats, especially in the context of IoT, has been achieved.

7.9 Open Challenges

Some work remains to be done in this research area. First, incorporating the CCAAC model with light-weight authentication supporting capability, such as the one proposed in [17], and to evaluate it with a more realistic adversaries model. Second, the current authority delegation framework in the proposed CCAAC assumes trust relationships that are already established among entities in different federation domain. The future outlook will consider the case in which no prior knowledge of the trust relationship between two network domains in Federated IoT. Unlike the approach used in this paper, an additional entity that is trusted by both domains, for instance Identity Provider (IdP), needs to be involved in the design. Lastly, prototype implementation of all aspects in CCAAC by using the existing access control realization model or on a device level will be something to look forward.

7.10 Conclusion

Access control is of paramount importance for a full thrive of IoT, especially due to the dynamic network topology and distributed nature. In this chapter, different access control models with their advantages and limitations have been discussed. Based on this, the secure CCAAC model with specifications of the capability (CAP) creation and access mechanisms has been proposed. The definition of Federated IoT is used as a baseline in designing the proposed authority delegation mechanism that is part of the overall CCAAC model. The protocol description and security consideration involving the usage of cryptographic keys are further presented in the paper to give some guidelines in the practical implementation. The proposed CCAAC has been analysed in the presence of security threats, such as eavesdropping and replay attack by an intruder, in order to test its resilience. Security proofs and evaluations by using AVISPA tool show that the CCAAC model achieves not only access control but also the secrecy of CAP and authentication.

Acronyms

CCAAC Capability-based Context Aware Access Control

IoT Internet of Things

FI Future Internet

M2M Machine-to-Machine

ITU-T International Telecommunication Union – Telecommunication Standardization Sector

WSN Wireless Sensor Network

RFID Radio Frequency Identification

NFC Near Field Communication

ICAP Identity based Capability

VID Virtual Identity

MAGNET My Adaptive Global NETwork

CASM Context Aware Security Manager

PN Personal Network

ACS Access Control Servers

PE Policy Engine

OS Object Service

SecaaS Security as a Service

ACL Access Control List

ACM Access Control Matrix

RBAC Role Based Access Control

XACML Extensible Access Control Markup Language

CWAC Context aWare Access Control

GTRBAC General Temporal RBAC

XML Extensible Markup Language

PTD Personal Trusted Device

WPAN Wireless Personal Area Network

MAC Message Authentication Code

DoS Denial of Service

PNDS PN Directory Service

CA Certificate Authority

FM Federation Manager

PN-F Personal Network Federation

SDP Security Decision Point

AVISPA Automated Validation of Internet Security Protocols and Applications

6LoWPAN IPv6 over Low-power Wireless Personal Area Network

IoT-DS IoT Directory Service

IoT-FM IoT Federation Manager

GW Gateway

IdP Identity Provider

PNDS Personal Network Directory Service

CARBAC Context Aware Role Based Access Control

CPFP Certified PN Formation Protocol

PFP PN Formation Protocol

PNCA Personal Network Certificate Authority

PKI Public Key Infrastructure

SSL Secure Socket Layer

TLS Transport Layer Security

DH Diffie–Hellman

ECC Elliptic Curve Cryptography

ECMQV Elliptic Curve Menezes-Qu-Vanstone

ECDH Elliptic Curve Diffie–Hellman

DHT Distributed Hash Table

P2P Peer-to-Peer

References

[1] Bayu Anggorojati, Parikshit N. Mahalle, Neeli R. Prasad, and Ramjee Prasad. Capability-based access control delegation model on the federated iot network. In *Proceedings 15th International Symposium on Wireless Personal Multimedia Communications (WPMC)*, September 2012.

[2] Tuomo Repo, Arto Hämäläinen, Jari Porras and Pekka Jäppinen. Applying wireless technology to access control systems. In *Proceedings 1st Workshop on Applications of Wireless Communications (WAWC'03)*, 2003.

[3] Tyler Close. ACLS don't. Technical report, Hewlett Packard Laboratories, 2009.

[4] D. Dolev and A. Yao. On the security of public key protocols. *IEEE Transactions on Information Theory*, 29(2):198–208, March 1983.

[5] Patrik Floréen, Michael Przybilski, Petteri Nurmi, Johan Koolwaaij, Anthony Tarlano, Matthias Wagner, Marko Luther, Fabien Bataille, Matthieu Boussard, Bernd Mrohs, and Sian Lun Lau. Towards a context management framework for mobilife. In *IST Mobile & Communications Summit*, 2005.

[6] H. Gomi. Dynamic identity delegation using access tokens in federated environments. In *Proceedings IEEE International Conference on Web Services (ICWS)*, July 2011.

[7] Hidehito Gomi, Makoto Hatakeyama, Shigeru Hosono, and Satoru Fujita. A delegation framework for federated identity management. In *Proceedings of the 2005 workshop on Digital Identity Management (DIM'05)*. ACM, New York, 2005.

[8] L. Gong. A secure identity-based capability system. In *Proceedings IEEE Symposium on Security and Privacy*, May 1989.

[9] M. Grossmann, M. Bauer, N. Honle, U.-P. Kappeler, D. Nicklas, and T. Schwarz. Efficiently managing context information for large-scale scenarios. In *Proceedings Third IEEE International Conference on Pervasive Computing and Communications*, March 2005.

[10] K. Hasebe and M. Mabuchi. Capability-role-based delegation in workflow systems. In *IEEE/IFIP 8th International Conference on Embedded and Ubiquitous Computing (EUC)*, Dec. 2010.

[11] Koji Hasebe, Mitsuhiro Mabuchi, and Akira Matsushita. Capability-based delegation model in RBAC. In *Proceedings of the 15th ACM Symposium on Access Ccontrol Models and Technologies (SACMAT'10)*. ACM, New York, 2010.

[12] http://avispa project.org/.

[13] Young-Gab Kim, Chang-Joo Mon, Dongwon Jeong, Jeong-Oog Lee, Chee-Yang Song, and Doo-Kwon Baik. Context-aware access control mechanism for ubiquitous applications. In *Advances in Web Intelligence*, Lecture Notes in Computer Science, volume 3528, pages 932–935. Springer, Berlin/Heidelberg, 2005.

[14] Devdatta Kulkarni and Anand Tripathi. Context-aware role-based access control in pervasive computing systems. In *Proceedings of the 13th ACM symposium on Access Control Models and Technologies (SACMAT'08)*. ACM, New York, 2008.

[15] Henry M. Levy. *Capability-Based Computer Systems*. Butterworth-Heinemann, 1984.

[16] Sue Long, Rob Kooper, Gregory D. Abowd, and Christopher G. Atkeson. Rapid prototyping of mobile context-aware applications: The cyberguide case study. In *Proceedings of the 2nd Annual International Conference on Mobile Computing and Networking (MobiCom'96)*. ACM, New York, 1996.

[17] Parikshit N. Mahalle, Bayu Anggorojati, Neeli R. Prasad, and Ramjee Prasad. Identity establishment and capability based access control (IECAC) scheme for Internet of Things. In *Proceedings 15th International Symposium on Wireless Personal Multimedia Communications (WPMC)*, September 2012.

[18] R. Prasad. *My Personal Adaptive Global NET (MAGNET)*. Signals and Communication Technology Book. Springer, the Netherlands, 2010.

8

Jamming and Physical Layer Security for Cooperative Wireless Communication

Vandana Rohokale, Neeli Prasad and Ramjee Prasad

Center for TeleInFrastuktur, Aalborg University, Denmark
e-mail: {vmr, np, prasad}@es.aau.dk

Abstract

Security functionality handled by higher layers of OSI model is a costly affair due to key management and complex mathematical encryption-decryption mechanisms in the scalable networks like wireless sensor networks (WSNs). WSN consists of resource constrained tiny sensor nodes for which battery is their lifetime. Physical layer security is the cost effective solution for such adhoc networks. The cooperative diversity mechanism makes use of the benefits of wireless sensor network scalability in terms of cooperative resource sharing in which multiple diversity channels are created which results into improved wireless connectivity. Cooperative jamming principle is used where the source transmits the encoded signal and at the same time relay transmits the weighted jamming signal to increase the equivocation of the eavesdropper. This work proposes a novel cooperative jamming mechanism for scalable networks like Wireless Sensor Networks (WSNs) which makes use of friendly interference to confuse the eavesdropper and increase its uncertainty about the source message. Channel fading is exploited to achieve perfect secrecy. The full communication link is built with the help of Information theoretic source and channel coding mechanisms. The whole idea is to make use of normally inactive relay nodes in the selective Decode and Forward cooperative communication and make them work as cooperative jamming sources to increase the equivocation of the eavesdropper. In

Fabrice Theoleyre and Ai-Chun Pang (Eds.), Internet of Things and
M2M Communications, 161–181.

this work, eavesdropper's channel's equivocation is compared with the main channel in terms of mutual information and secrecy capacity.

Keywords: wireless sensor networks (WSNs), cooperative jamming, eavesdropper's equivocation, mutual information, secrecy capacity.

8.1 Introduction

The performance of wireless networks is greatly affected by some of the channel parameters such as bandwidth and power scarcity, multi-user interference, non-reliability due to signal fading, vulnerability to the attacks, etc. The cooperative diversity mechanism makes use of the benefits of wireless sensor network scalability in terms of cooperative resource sharing in which multiple diversity channels are created which results into the higher transmission rates, increased throughput and coverage range, improvement in reliability and end-to-end performance and much more. Cooperative wireless communication (CWC) greatly improves the cross layer optimizations.

Wireless sensor nodes are inherently memory and energy constrained. Today's commonly utilized algorithms such as RSA, Diffie–Hellman, NTRU and Elliptic Curve Cryptography make use of large numbers multiplication in their encryption and decryption mechanisms. Due to their huge demand of memory and energy, these cryptographic algorithms cannot be employed to wireless sensor nodes. This research work proposes a cooperative jamming technique for physical layer security with the help of Information Theoretic source and channel coding mechanisms. In CWC, the active nodes may increase their effective QoS via cooperation. Out of three most popular cooperative relaying strategies, amplify and forward mechanism results in the noise amplification and is not suitable for the scalable networks like wireless sensor networks. The Decode-and-forward mechanism is well suited for WSN provided that the channels are strong enough. Cooperative jamming is designed to mystify the eavesdropper.

Wireless sensor nodes are resource constrained miniature devices and energy reservation for lifetime extension is crucial for them. Most of the energy consumption occurs due to the transmissions. Hence, reduction in the amount of data transmissions can result in energy savings. Information theoretic communication concept was born after the evolutionary paper by Cloud Shannon entitled "A Mathematical Theory of Communication". Physical layer security and information theory are closely related to each other and are extensively studied in [1], which opened the doors for the security considerations at physical layer. WSN is a low data rate system.

To limit the data rates, efficient compression is necessary. Various source coding techniques are available in the literature including Shannon–Fano coding, Huffman coding, Lempel Ziv Welch coding, Arithmetic coding, Adaptive coding mechanisms, Run length encoding, etc. for effective data compression. Channel coding mechanisms like LDPC, Convolution coding, BCH coding and Reed Solomon are proven to be reliable which can provide the lower values of BER rates with inbuilt error detection and correction capability. Coded modulation is a concept which combines channel coding and modulation in a single block providing the added benefit of bandwidth efficiency without any change in the symbol rate and power spectrum [2].

The chapter is organized as follows. Section 8.2 elaborates on the related work in the field of cooperative wireless communication and physical layer security. Proposed secure CWC model is depicted in Section 8.3, which includes the details about the source coding, channel coding and modulation. Section 8.4 discusses the simulation results explaining the secrecy capacity of main and eavesdropper's channels. Finally, we conclude the chapter in Section 8.5 with discussions on the future scope for this work.

8.2 Related Work

In wireless information, transmitted information is also overheard by the neighboring nodes other than the intended recipient node. Cooperative wireless communication (CWC) makes use of broadcast nature of the wireless networks in including most of the neighboring receiver nodes in the relaying mechanism. Cooperative broadcasting has several advantages as compared with the multi-hop broadcasting in terms of improved coverage ranges and connectivity, better power efficiency and much better communication data rates. CWC forms the situation of virtual multiple input multiple output (MIMO) by making use of the group relaying node antennas. Traditional multi-hop networks tend to generate the contention in the traffic whereas the cooperative transmission ensures the traffic regulation.

Relaying and cooperative diversity essentially creates a virtual antenna array. All of the cooperative diversity protocols are efficient in terms of full diversity achievement and optimum performance except fixed decode-and-forward approach. Although prior to Laneman [3], the work on relay and cooperative channels utilized full duplex approach, he has constrained the cooperative communication to employ half duplex transmissions. Also in this case, the Channel State Information (CSI) is employed in the receiver instead of transmitter. Many research works have investigated the advantages of cooperative transmission over traditional multi-hop communication. Cooperative opportunistic large array approach is proposed in [4], wherein, the network nodes transmit the overheard information based on their stored energy in the distributed manner. As compared to centralized technique, this method is less complex since it eliminates the problem of transmission scheduling.

The authors have explored energy efficiency of CWC over multi-hop networks by considering different setups in [4–6]. They have taken into account transmit diversity and receiver combining mechanism. Also for cooperative transmission, they have considered pre-scheduling with power allocation policy to minimize the total power consumption in the network. In the research works of [7, 8], different cooperative strategies are proposed for improvement in the network lifetime, network coverage and communication data rates. Authors have investigated the broadcast capacity of the cooperative wireless networks for slowly fading channels in [9]. They have considered a model in which the outage is declared if any of the in between relay receivers fail to decode the source message. Also, they have defined broadcast capacity in terms of maximum data rate at which the outage probability converges to zero as the number of cooperating nodes in the network reaches infinity.

Wire-tap channel was introduced by Wyner where the eavesdropper's channel is assumed to be degraded as compared to the legitimate receiver's channel. The positive perfect secrecy was achieved for this single user wiretap channel [10]. Csiszar and Korner [11] studied the single user eavesdropping channel which was not necessarily degraded and obtained the secrecy capacity. They introduced superposition coding technique for the broadcast channels. Physical layer security intends to make use of the inherent randomness in the wireless channels to provide the additional security at physical layer. Statistical independence between the eavesdropper's observation and the actual message is an important measure for the Information Theoretic Security (ITS). ITS is measured by Eve's uncertainty about the message given the code word, called Eve's equivocation [12]. According to Maurer [13], for cryptographic causes the noisy communication channel should be converted into an error free channel by combining the cryptographic coding with the error control coding. Also, mere difference in between the signals received by an eavesdropper and the intended recipient may be sufficient for the achievement of cryptographic security.

As compared to the conventional cryptosystems, the secrecy assured by the Information theoretic mechanisms is more cost effective solution because it avoids the key generation and management tasks and results into the significantly lower complex solutions which are proven to provide the savings in the resources like memory and battery supply which are the critical issues for the resource constrained wireless sensor networks. Also the information theoretic security techniques are less prone to the man-in-the-middle attack as compared with the public key cryptosystems because of the inherent randomness shared by the communicating radio nodes [14–18]. The ITS mechanisms are shown to be vital for the dominant eavesdroppers which possess unlimited computational resources and have access to the communication systems either through perfect or noisy channels.

Bloch and Barros in their work [19] have shown adorable results with the appropriate combination of Information Theoretic Security and Cryptography. They have shown that source and channel coding techniques with small cryptographic one time pads can achieve perfect secrecy. The authors have proposed a key pre-distribution scheme which makes use of a mobile node for the task completion of key distribution process blindly using network coding techniques. This methodology has shown to reduce the memory requirements. For secure communication, cryptography is not the only solution. By exploiting the intrinsic randomness of the channels and state-of-the-art error correcting codes, we can implement reliable and insensible data transfer which is the building block of the secure multi-node communication.

Research work by Liang Chen [20] shows that the physical layer security can be achieved even though the relays possess lower security clearance values when the compress and forward cooperating mechanism is used. The authors propose the combined version of decode and forward with compress and forward for ripping the benefits from the advantages of both of them. They propose that when the channels are better and the relay nodes have higher security clearance figures then decode and forward scheme works better and for other conditions compress and forward scheme is the best choice. With these cases, the high transmission rates and physical layer security can be achieved.

In [21] physical layer security issues for wireless communication are discussed in a tutorial fashion. The secret channel capacity and computational capacity are considered as the metrics for the comparison among different physical layer tactics. The authors have classified the existing physical layer security techniques into five major categories based on their characteristic features. These classes can be listed as: theoretical secure capacity, power, code, channel and signal detection methodologies. Shuangyu Luo et al. [22] proposed an optimally organized Gaussian noise for cooperative jamming which results in the maximum secrecy rate. The authors presented that, when the optimal solution requires global channel information, the suboptimal solution requires only local channel information. The suboptimal solution almost reaches close to the optimal solution for the achievement of secrecy rate. Uncoordinated cooperative jamming mechanism is proposed wherein no eavesdropper channel information is needed for the secure communication.

A cooperative jamming protocol system design for secure wireless communication with the consideration of a relay with multiple antennas is proposed in [23] for determining the antenna weights and transmit power of source and relay, so that the system secrecy rate is maximized with the considerable decrease in the transmit power. Relay equipped with multiple antenna can provide degrees of freedom for the relay channel and thus can exclude the effects of jamming at the receiving end. In [24], the authors proposed a novel full cooperative jamming and partial cooperative jamming methodologies for two hop decode and forward wireless MIMO relay systems in the presence of the eavesdropper. Full cooperative jamming and partial cooperative jamming depends on whether the transmitter and helper are transmitting the jamming signals at the same time. With these proposed cooperative jamming schemes, the source and destination nodes act as momentary helpers for transmission of the jamming signals during their inactive phases.

In [25], the authors presented a new physical layer approach called iJam for secret key generation which they claim to be fast and channel independent. The iJam mechanism works as follows. The transmitter sends its transmission twice. The receiver works as the jammer too. The receiver cum jammer jams gratis samples from the original transmitted signal and its repeated version. At the time of decoding at the receiving end, the receiver combines together the unjammed samples from both the received signals to reconstruct the original exact transmitted signal. The working principle behind the iJam technique is that receiver jams the transmissions so that the information about the undisclosed key is kept secret from the eavesdropper while the intended recipient is allowed to extract the secret key perfectly by stitching the unjammed received symbols.

In [26], the use of cooperating relays for the performance improvement of secure wireless communication with the eavesdropper's presence is proposed. The authors have taken into consideration three cooperative protocols namely decode-and-forward, amplify-and-forward and cooperative jamming. With these cooperative protocols, two practical problems like transmit power

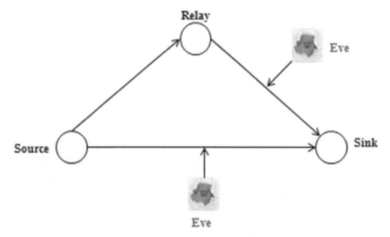

Figure 8.1 Simplified cooperative relay model with eavesdropper.

allocation at the source and relays and relay weight determination for the achievable secrecy rates are analyzed. The improvement is observed in the limitations of channel conditions and the overall system performance with cooperation as compared to without cooperation strategies.

This research work proposes joint source and channel coding mechanism for the security in cooperative wireless communications. A Lempel–Ziv–Welch (LZW) source coding technique is used for secure data compression which will result in the low data rates and ultimately low energy consumptions. BCH channel coding method is utilized to achieve the appropriate reliability values. For modulation, a DSSS modulation with Gold codes is considered which proves to be an inherently secure modulation technique. Coded modulation is proposed, which proves to be a bandwidth efficient solution for the miniature WSN entities.

8.3 Proposed Reliable and Secure Cooperative System

The simplified cooperative relaying mechanism is depicted in Figure 8.1 with only one relay and two eavesdroppers, one on the main channel and the other on the relay channel. In the first time slot, source transmits the message to relay as well to the sink. In the same time slot, the relay will transmit the corresponding jamming signal intended towards the eavesdropper. The complete functional block diagram is shown in Figure 8.2. It includes the source coding (decoding), channel coding (decoding) and the modulation (demodulation).

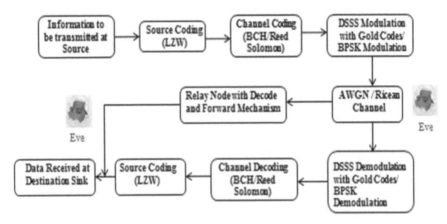

Figure 8.2 Block diagram of the proposed information theoretically secure CWC.

An information theoretic mechanism with guaranteed secrecy has the added advantage of relief from key management issues which results into considerable resource savings as compared to traditional cryptosystems. As compared to the public key cryptosystems, information theoretic security approaches are found to be less prone to man-in-the-middle attack due to the inherent randomness shared by the nodes [12]. Since the cooperative wireless sensor network nodes are battery limited devices, energy is the crucial parameter for their lifetime. Most of the energy consumption in wireless communications is due to the radio communication among the node entities. For limiting transmission data rates, the data compression mechanism is proved to be an energy efficient tool [27]. Out of all the available data compression algorithms, the LZW source coding mechanism seems to be inherently secure.

The Lempel–Ziv technique can be modified to provide proper authentication requirement by the CWC for sensor networks. Good compression ratios ultimately result in considerable energy savings. BCH channel coding and decoding mechanism is utilized here. Light weight version of the LZW and BCH coding will be utilized here for cooperative WSN. Cooperative jamming mechanism is implemented here. When the source transmits the actual information, at the same time, relay sends random jamming signal. Jamming signal received at the receiver is cancelled by making use of Interference cancellation technique. And the jamming signal received by the eavesdropper is treated as the actual transmitted signal and it tries to decode it. After successful implementation of the block diagram as shown in Figure 8.2, many simulations are performed for this case and then the probabilistic entropies, mutual information and channel secrecy capacities are calculated.

The concept of coded modulation is introduced for the physical layer security in the cooperative communication as depicted in Figure 3. The coded communication combines the channel coding and modulation blocks into a single block. The evolution from the traditional cryptographic communication link towards the bandwidth efficient coded modulation inclusive communication link through the information theoretic secured communication link is depicted in this figure. Encryption for security and channel coding for reliability blocks from the conventional cryptographic system are combined in a single secure encoding block in the information theoretic communication link. Secure encoding and modulation blocks in the information theoretic communication system are shown to be further combined in a single block named coded modulation encoder in the final communication link. This evolution has provided users with numerous benefits in terms of energy efficiency, reliability, security and bandwidth efficiency.

The following parameters are taken into consideration:

1. Entropy of the system as a whole $H(X, Y)$ – average information per pairs of transmitted and received characters.

$$H(x, y) = \sum_{i=1}^{n} \sum_{j=1}^{m} p(x_i, y_j) \log \frac{1}{p(x_i, y_j)} \tag{8.1}$$

2. Entropy of the source $H(X)$ – average information per character of the source.

$$H(x) = \sum_{i=1}^{n} p(x_i) \log \frac{1}{p(x_i)} \tag{8.2}$$

3. Entropy at the receiver $H(Y)$ – average information per character at the destination.

$$H(y) = \sum_{j=1}^{m} p(y_j) \log \frac{1}{p(y_j)} \tag{8.3}$$

4. Conditional Entropy $H(Y|X)$ – a specific character x_k being transmitted and one of the permissible y_j may be received (a measure of information about the receiver, where it is known what was transmitted). $H(Y|X)$ gives an indication of the noise (errors) in the channel.

$$H(y|x) = H(x; y) - H(x) \tag{8.4}$$

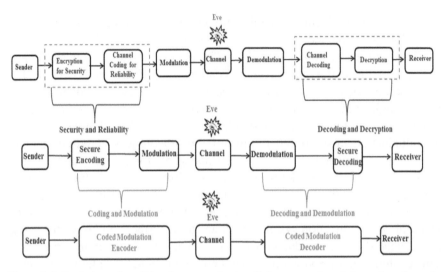

Figure 8.3 Evolution of traditional communication link to a modern information theoretic communication link.

5. Conditional Entropy $H(X|Y)$ – a specific character y_j being received; this may be the result of transmission of one of the x_k with given probability (a measure of information about the source, where it is known what was received). $H(X|Y)$ gives a measure of equivocation (how well one can recover the input content from the output).

$$H(x|y) = H(x; y) - H(y) \qquad (8.5)$$

The mutual information for main channel is given by

$$I(x; y) = H(x) - H(x|y) \qquad (8.6)$$

6. Similarly, mutual information for the eavesdropper's channel is given by

$$I(x; e) = H(x) - H(x|e) \qquad (8.7)$$

7. The maximum amount of mutual information is nothing but the secrecy capacity for that particular channel.

$$C_{SM} = \max[I(x; y)] \qquad (8.8)$$

Ultimate aim was to prove that $H_{E/X} > H_{Y/X} > H_{R/X}$ and $C_{sE} < C_{sM} < C_{sR}$, where C_{sE} is the secrecy capacity of the eavesdropper's channel; C_{sM}

is the secrecy capacity of the main or direct (main) channel; and C_{sR} is the secrecy capacity of the relay channel. The maximum amount of eavesdropper's equivocation (uncertainty of eavesdropper about the source message) indicates the system security.

8.3.1 LZW Source Coding and Decoding

LZW is the most popular dictionary based lossless data compression technique. LZW is the algorithm where Welch has added some modifications to the prior available methodologies by Lempel and Ziv known as LZ77 and LZ78 [28]. As compared to the Shannon-Fano and Huffman coding techniques, the LZW mechanism provides better compression ratios with improved coding efficiency. LZW source encoding is skilled by parsing the source data sequence into the fragments that are the shortest sub sequences not encountered previously [29]. Let us consider one example data sequence to illustrate this algorithm as follows:

$$0100111110010100000101010101100110000$$

The binary symbols 0 and 1 are assumed to be dictionary bits which are assumed to be codebook. So, we get

Codebook/Dictionary Bits: 0, 1

Data to be fragmented:

$$0100111110010100000101010101100110000$$

And then the next set can be

Decoded Data with Dictionary Bits:

$$01, 00, 11, 111, 001, 010, 000, 0101, 01011, 0011, 0000$$

8.3.2 BCH Channel Encoding and Decoding

BCH codes form the subclass of cyclic codes which are powerful random error correcting codes. It is the good generalization of the Hamming codes for multiple error correction. BCH codes were simplified by making use of Galois field and primitive polynomials by Gorenstein and Zierler [30]. Berlekamp's iterative algorithm and Chien's search algorithm are the most

Table 8.1 LZW algorithm for sequence 0100111110010100000101010110011000.

Numerical Positions	Subsequence (Parsed Data)	Codebook (Numerical Representation)	Binary Encoded Blocks	Decoded Data (with dictionary bits numerical position in brackets)
1	**0**			**0** (1)
2	**1**			**1** (2)
3	01	12	00011	01 (3)
4	00	11	00010	00 (4)
5	11	22	00101	11 (5)
6	111	52	01011	111 (6)
7	001	42	01001	001 (7)
8	010	31	00110	010 (8)
9	000	41	01000	000 (9)
10 (A)	0101	82	10001	0101 (A)
11 (B)	01011	A2	10101	01011 (B)
12 (C)	0011	72	01111	0011 (C)
13 (D)	0000	91	10010	0000 (D)

efficient BCH decoding algorithms. The most common binary BCH codes are known as primitive BCH codes and are characterized as follows:

$$\text{Block Length: } n = 2^m - 1$$

$$\text{Number of message bits: } k \geq n - mt$$

$$\text{Minimum Distance: } d_{\min} \geq 2t + 1$$

Each BCH code is a t error correcting code which can detect and correct up to t random errors per codeword. Let α be a primitive element of $GF(2^m)$. The generator polynomial $g(x)$ of t error correcting BCH code of length $2^m - 1$ is the lowest degree polynomial over $GF(2)$ which has $\alpha, \alpha^2, \alpha^3 \ldots \alpha^{2t}$ as its roots. Let $f_i(x)$ be the minimal polynomial of α^i. Then $g(x)$ must be the LCM of $f_i(x)$ as follows:

$$g(x) = LCM\{f_1(x), f_2(x), \ldots, f_{2t}(x)\} \tag{8.9}$$

8.3.3 Reed Solomon Channel Encoding and Decoding

Reed Solomon codes are an important subclass of the non-binary BCH codes with a wide range of applications in mobile communication and data storage. Reed and Solomon in their paper [31], introduced the idea of burst error correcting RS codes. In this subclass of BCH codes, the symbol field (subfield) $GF(q)$ and the error locator field $GF(q^m)$ are the same, i.e., $m = 1$.

Hence, for this case,

$$n = q^m - 1 = q - 1 \tag{8.10}$$

The minimal polynomial of any element β in the same field $GF(q)$ is

$$f_{\beta(x)} = x - \beta \tag{8.11}$$

Since the symbol field and error locator field are the same, all the minimal polynomials are linear. The generator polynomial for a t error correcting code will be given by

$$g(x) = LCM[f_1(x) f_2(x), \ldots, f_{2t}(x)] \tag{8.12}$$

$$= (x - \alpha)(x - \alpha^2) \ldots (x - \alpha^{2t-1})(x - \alpha^{2t}) \tag{8.13}$$

Hence the degree of the generator polynomial will always be $2t$. Thus, the RS code satisfies $n - k = 2t$.

8.3.4 DSSS Modulation and Demodulation with Gold Codes

For an indoor scenario, DSSS is the best choice. Gold codes are the types of binary sequences which use a number of PN sequences to increase the security in WSN [32]. The noisy nature of the wireless medium is exploited to improve the security of overall communication system. Gold codes have restricted small cross correlations within a set, which is useful when multiple devices are broadcasting in the same range. The steps for Gold code generation are as follows:

- Choose two maximum length sequences of the same length $2^k - 1$ such that cross correlation is less than or equal to $2^{(k+2)/2}$, where k is the size of the LFSR used to generate the maximum length sequence.
- The set of the $2^k - 1$ EX-ORs of the two sequences in their various phases is a set of Gold codes. The highest absolute cross-correlation in this set of codes is $2^{(k+2)/2} + 1$ for even k and $2^{(k+1)/2} + 1$ for odd k. The EX-OR of two Gold codes from the same set is another Gold code in some phase.

8.3.5 Coded Modulation and Spectral Efficiency

Conventionally, the binary coding and modulation are considered to be two separate techniques in which we send R information bits/symbol and achieve spectral efficiency R. Constant transmission rate demands for the bandwidth expansion by a factor of $1/R$. Until Ungerboeck [33], the channel coding was not considered useful for achieving spectral efficiency or coding gain is achieved at the expense of bandwidth expansion. Coded modulation is a technique in which there is concatenation in between error correcting codes and the signal constellation technique. Here, groups of coded bits are mapped into points in the signal constellation in such a way as to increase the Euclidian distance in between the signal points. There are three major types of coded modulation:

- Trellis Coded Modulation (TCM),
- Block Coded Modulation (BCM),
- Turbo Coded Modulation.

The concept of Trellis Coded Modulation is illustrated in Figure 8.4. The functions of a TCM consist of a Trellis code and a constellation mapper as shown in this figure. TCM combines the functions of a convolutional coder of rate $R = k/k + 1$ and a M-ary signal mapper that maps $M = 2k$ input points into a larger constellation of $M = 2k + 1$ constellation points [34]. According to Information Theory, for optimal communication, long sequences of signals should be designed with maximum separation among them and at the receiving end; decision making should be performed on such lengthy signals rather than on bits or symbols. Trellis and multi-dimensional codes are designed to maximize the Euclidian distance between possible sequences of transmitted symbols. The larger the Euclidian distance, the lower the bit error rate making the communication more reliable.

TCM is a bandwidth efficient modulation based on convolution coding and Viterbi decoding. It safeguards bandwidth by doubling the number of constellation points of the signal. Because of this, the bit rate increases but the symbol rate remains same. In multi-dimensional TCM, the number of symbols created in one processing period is increased. Transmitted symbols are generated together and this co-generation creates dependence and allows better performance. Instead of the effective code rate being 2/3 as in 1×8 PSK, it can be higher, e.g. 5/6, 8/9 or 11/12 in multi-dimensional TCM.

8.4 Simulation Results

The total communication link is prepared with Matlab codes. For simulation purpose, the sequence of 300 bits is randomly generated and applied as an input to the LZW source coding block. The input to the BCH coding block is the binary sequence of 448 bits. Then (511, 448) BCH coding is applied with the error correcting capability of seven bits. The output of BCH coding

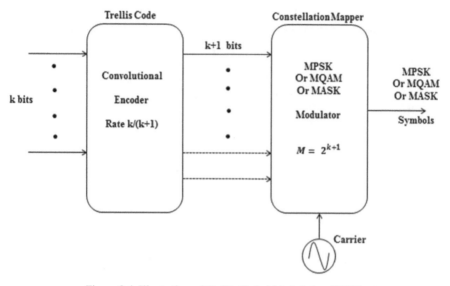

Figure 8.4 Illustration of Trellis Coded Modulation (TCM).

Table 8.2 Communication link design parameters.

Parameter	Technique Used	No. of input bits	No. of output bits
Input Data sequence	Binary data	300	
Source Coding	LZW Coding	300	448
Channel Coding	BCH Coding	448	511
(n, k, t)	(511, 448, 7)		
Modulation	DSSS with	Gold code	10220
	Gold Codes	length = 20 bits	
Channel	AWGN channel	10220	10220
	with noise addition		
Demodulation	DSSS Demodulation	10220	511
Channel Decoding	BCH Decoding	511	448
Source Decoding	LZW Decoding	448	300

block is containing 511 bits which is the input to DSSS modulation block. Twenty bit lengthy PN sequence is used for the construction of the Gold Codes. Accordingly the 511 bits data is spreaded to 10220 bits after the DSSS modulation. The AWGN channel is taken into consideration for this experimentation. Table 8.2 gives detail information about the techniques used in building the communication link.

Figure 8.5 shows the BCH encoded data, its spreaded and despreaded version and data recovered after DSSS demodulation. BCH encoded data and

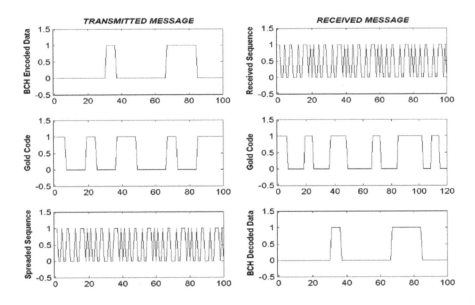

Figure 8.5 BCH encoded input data, spreaded sequence, despreaded sequence and recovered data after demodulation.

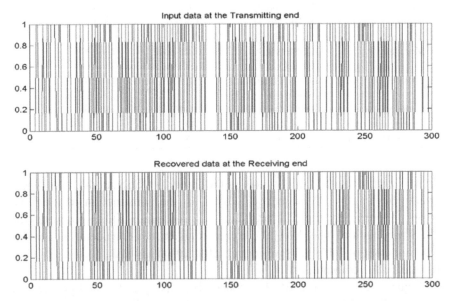

Figure 8.6 Source input data and recovered data at the receiving end.

BCH decoded data look exactly same. Information to be transmitted at the source input and the recovered data at the final receiver or sink block are shown in Figure 8.6. The original data is successfully recovered at the receiving end. First part of Figure 8.6 is the input to the LZW source coding block and the second part of Figure 8.6 is the output of the LZW source decoding block. We observe the exactly same input and output bit sequences at the source coding input and the source coding output blocks. This indicates zero error probability which is the indication of maximum mutual information and ultimately it results into the maximum secrecy capacity.

If the probability of error is zero, then the data rate at which the transmitter communicates to the receiver is given by

$$\text{Date Rate} = \frac{\log_2 (\text{Number of Messages})}{n} = \frac{nR}{n} = R \text{ bits/sec} \qquad (8.14)$$

Then the channel capacity becomes

$$C = \max I(x; y) \qquad (8.15)$$

Shannon states that "What can be achieved by the best strategy over n channel uses is given by the maximal mutual information for a single channel use" [2]. Secrecy capacity of a communication channel is nothing but the maximum amount of mutual information shared between source and receiver entities.

As is clear from Figure 8.7, mutual information of the main (source-receiver) channel $I(X; Y)$ is much higher than the mutual information of the eavesdropper's channel $I(X; E)$. This shows that the secrecy capacity of the main channel is greater than the secrecy capacity of the eavesdropper's channel due to the jamming effect by the cooperating relay node. Hence, the eavesdropper's equivocation is much higher than the main channel. This is really encouraging factor for the physical layer security. Also, the source and channel coding mechanisms taken into consideration are likely to consume less resources as expected by wireless sensor networks.

As is depicted in Figure 8.7, secrecy capacity of the main channel is found to be almost equal to one. It is in the range of 0.96 to 0.989. At the same time, the secrecy capacity of the eavesdropper's channel is found to be much lower as compared to the main channel and it ranges from 0.005 to 0.0087. This indicates the security of the communication link under consideration.

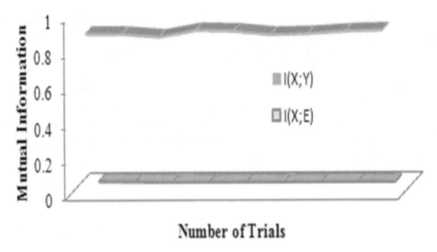

Figure 8.7 Secrecy capacities of the main and eavesdropper's channel in terms of mutual information.

8.5 Conclusions

The maximum amount of uncertainty for the eavesdropper's channel is the indication of the security of the communication system. Also, secrecy capacity of the main communicating channel is found to be greater than the eavesdropper's channel which is the promising result for further work. The secrecy capacity of the eavesdropper's channel is found to be much lower as compared to the main channel and it ranges from 0.005 to 0.0087. The secrecy capacity of the main channel is found to be in the range of 0.96 to 0.989. Perfect secrecy is achieved with the help of cooperative jamming and information theoretic source and channel coding techniques.

8.6 Future Research Directions

This work can be extended further for different cooperative relaying techniques such as decode and forward and amplify and forward while taking into consideration of weights and polarization at the relays. It can also be extended further for analysis of the spectral efficiency and energy efficiency for cooperative communication with information theoretic security. In cognitive radio networks, for the spectrum management and security issues, this research work can be extended further with consideration of different vulnerabilities in the system. Researchers can experiment by making use of different source coding techniques like Huffman, Shannon Fano, Arithmetic, etc., channel coding mechanisms like BCH, RS, Convolution, TCM, etc., and different modulation techniques depending on particular application. This communication link can also be built by making use of channel assumptions such as Ricean, Rayleigh or Broadcast channels.

References

[1] C. E. Shannon. Communication theory of secrecy systems. *The Bell System Technical Journal*, 28:656–715, October, 1949.

[2] V. M. Rohokale, N. R. Prasad, and R. Prasad, Cooperative jamming for physical layer security in wireless sensor networks. In *Proceedings of 15th International Symposium on Wireless Personal Multimedia Communications*, Taipei, Taiwan, September 2012.

[3] A. Scaglione and Y.-W. Hong. Opportunistic large arrays: Cooperative transmission in wireless multihop ad-hoc networks to reach far distances. *IEEE Transactions on Signal Processing*, 51(8), August 2003.

[4] Y.-W. Hong and A. Scaglione. Energy-efficient broadcasting with cooperative transmission in wireless sensory ad hoc networks. In *Proceedings of the Allerton Conference on Communication, Control and Computation (ALLERTON)*, October 2003.

[5] I. Maric and R. D. Yates. Cooperative multihop broadcast for wireless networks. *IEEE Journal on Selected Areas of Communication*, 22(6), August 2004.

[6] Y.-W. Hong and A. Scaglione. Energy-efficient broadcasting with cooperative transmission in wireless ad hoc networks. *IEEE Transactions on Wireless Communications*, 5(10), 2844–2855, October 2006.

[7] E. Sirkeci-Mergen and A. Scaglione. Coverage analysis of cooperative broadcast in wireless networks. In *Proceedings of IEEE Workshop on Signal Processing, Advances in Wireless Commun. (SPAWC)*, July 2004.

[8] B. Sirkeci-Mergen and A. Scaglione. A continuum approach to dense wireless networks with cooperation. In *Proceedings of Annual Joint Conference of the IEEE Computer and Communication Societies (Infocom)*, August 2005.

[9] A. Khisti, U. Erez, and G. Wornell. Fundamental limits and scaling behavior of cooperative multicasting in wireless networks. *IEEE Transactions on Information Theory*, 52(6), 2762–2770, June 2006.

[10] A. Wyner. The wire-tap channel. *Bell System Technical Journal*, 54:1355–1387, 1975.

[11] I. Csiszar and J. Korner. Broadcast channels with confidential messages. *IEEE Transactions on Information Theory*, IT-24(3):339–348, May 1978.

[12] Y. Liang, H. V. Poor, and S. Shamai. Information theoretic security. *Foundations and Trends in Communications and Information Theory*, 5, 2008.

[13] U. M. Maurer. Perfect cryptographic security from partially independent channels. In *Proceedings of the Twenty-Third Annual ACM Symposium on Theory of Computing (STOC'91)*, pages 561–571, January 1991.

[14] L. Lai, H. El Gamal, and H. V. Poor. Authentication over noisy channels. *IEEE Transactions on Information Theory*, 55:906–916, February 2009.

[15] J. L. Massey. Contemporary cryptography – An introduction. In *Contemporary Cryptography – The Science of Information Integrity*. IEEE, Piscataway, NJ, 1992.

[16] U. M. Maurer. Authentication theory and hypothesis testing. *IEEE Transactions on Information Theory*, 46:1350–1356, July 2000.

[17] U. Rosenbaum. A lower bound on authentication after having observed a sequence of messages. *Journal of Cryptology*, 6(3):135–156, 1993.

[18] G. J. Simmons. Authentication theory/coding theory. In *Proceedings of the CRYPTO'84 on Advances in Cryptography*, Lecture Notes in Computer Science, pages 411–431. Springer, New York, 1985.

[19] M. Bloch and J. Barros. *Physical Layer Security: From Information Theory to Security Engineering*. Cambridge University Press, 2011.

[20] Liang Chen. Physical layer security for cooperative relaying in broadcast networks. In *Proceedings Military Communications Conference (MILCOM 2011)*, USA, pages 91–96, 2011.

[21] Yi-Sheng Shiu, Shih Yu Chang, Hsiao-Chun Wu, S. C.-H. Huang, and Hsiao-Hwa Chen. Physical layer security in wireless networks: A tutorial. *IEEE Wireless Communications Journals and Magazines*, 18(2):66–74, April 2011.

[22] S. Luo, J. Li, and A. Petropulu. Physical layer security with uncoordinated helpers implementing cooperative jamming. In *Proceedings 7th IEEE Sensor Array and Multichannel Signal Processing Workshop (SAM2012)*, Hoboken, NJ, June 2012.

[23] Zhu Han, A. P. Petropulu, and H. V. Poor. Cooperative jamming for wireless physical layer security. *Proceedings IEEE/SP 15th Workshop on Statistical Signal Processing (SSP'09)*, pages 417–420, 2009.

[24] Jing Huang. Cooperative jamming for secure communications in MIMO relay networks. *IEEE Transactions on Signal Processing*, 59(10):4871–4884, October, 2011.

[25] Shyamnath Gollakota and Dina Katabi. Physical layer wireless security made fast and channel independent. In *Proceedings of IEEE Conference (INFOCOM)*, pages 1125–1133, April 2011.

[26] Lun Dong, Zhu Han, Athina P. Petropulu, and H. Vincent Poor. Improving wireless physical layer security via cooperating relays. *IEEE Transactions on Signal Processing*, 58(3):1875–1888, March 2010.

[27] Francesco Marcelloni and Massimo Vecchio. An efficient lossless compression algorithm for tiny nodes of monitoring wireless sensor networks. *The Computer Journal Advance Access*, April 2009.

[28] T. A. Welch. A technique for high-performance data compression. *Computer*, 17(6):8–19, June 1984.

[29] S. Haykin. Fundamental limits in information theory. In *Communication Systems*, 4th Edition. Wiley Publications, 2001.

[30] R. H. Morelos-Zaragoza and S. Lin. On primitive BCH codes with unequal error protection capabilities. *IEEE Transactions on Information Theory*, 41(3):788–790, May 1995.

[31] I. S. Reed and G. Solomon. Polynomial codes over certain finite fields. *Journal of the Society for Industrial and Applied Mathematics*, 8(2):300–304, June 1960.

[32] M. George, M. Hamid, and A. Miller. Gold code generators in virtex devices. XILINX Application Note: Virtex Series, Virtex-II Series, and Spartan-II family, XAPP217 (v1. 1), 10 January 2001.

[33] G. Ungerboeck. Channel coding with multilevel or multiphase signals. *IEEE Transactions on Information Theory*, IT-28:55–67, January 1982.

[34] A. J. Viterbi, J. K. Wolf, E. Zehavi, and R. Padovani. A pragmatic approach to Trellis Coded Modulation. *IEEE Communications Magazine*, pp. 11–19, July 1989.

9

Performance Modeling and Simulation of Machine-to-Machine (M2M) Systems*

Shih-Hao Hung[1], Chun-Han Chen[1] and Chia-Heng Tu[2]

[1]*National Taiwan University, Taipei, Taiwan*
[2]*Institute for Information Industry, Taipei, Taiwan*
e-mail: {hungsh, r99944045}@csie.ntu.edu.tw, chiahengtu@iii.org.tw

Abstract

Machine-to-machine (M2M) communications have been recently employed in various application domains, including smart homes, surveillance, remote control, healthcare, etc., where heterogeneous devices interact with each other via heterogeneous networks. The architecture of an M2M system is critical to its cost and performance, especially for those applications with strict real-time requirements. In addition, energy consumption is important to those devices which are powered by batteries. Considering the heterogeneity of devices and networks, the complexity for developing and evaluating the cost, performance and energy consumption of an M2M system can be very challenging, as it requires the developer to deal with issues which do not exist in conventional systems.

This chapter introduces a framework for evaluating the performance of M2M systems via simulation. The framework enables the user to quickly model an M2M system by running the M2M application over virtual machines and virtual network devices. The timing models in the virtual machines and a virtual network devices are design parameters taken from the actual system. The results of the simulation reveal the details of execution and es-

*This work was supported by grants from the National Science Council (101-2221-E-002-053- and 102-2218-E-007-004-).

timate the energy consumption for the M2M system. As illustrated in the case studies in this chapter, with such a tool, the developer may explore the design space to search for cost-effective or energy-efficient designs.

Keywords: M2M communication, performance estimation, virtual machine.

9.1 Introduction

With the advance of wire and wireless network technology, *machine-to-machine* (M2M) communications have widely been used to create sophisticated systems and applications with improved productivity. Example applications are goods tracking, building automation, elder health care, and smart home. In a simple case, a sensor captures an event and relays the event through a network to a machine which has a higher computational or storage capability to process the captured event. In general, M2M communications can be facilitated by wired or wireless networks to allow different types of devices to collaborate and accomplish sophisticated applications.

Depending on the application, an M2M system may span over thousands of devices that are of the same or different types. Based on the computational and storage requirements of the application, various types of machines and communication networks could be considered and selected during the design of an M2M system. Unlike a *wireless sensor networks* (WSN), the hardware devices in an M2M system are usually more diversified, and the characteristics of the machines may vary greatly, ranging from low-end sensors powered by micro-controllers to highly capable human-centric appliances. The software running on the machines can also be far more sophisticated.

The networks used to connect machines are also diversified. There are devices connected via short-range wireless sensor networks, body-area network, or personal wireless networks using standards such as ZigBee and Bluetooth. Beyond that, there are wired and wireless local area networks and wide-area networks using Ethernet, WiFi, 3G, 4G, and satellite networks. Sometimes, cloud-based computational or storage services may be involved, and the M2M communications can go through multiple heterogeneous networks.

Performance evaluation is important to the design of a system. Especially for an M2M system, rapid and accurate performance evaluation prior to its deployment is often highly valuable in practice, since it would incur a high cost to change the design after its deployment. Unfortunately, with the various levels of heterogeneity in software, hardware and network, it can be very complicated to evaluate an M2M system. In this chapter, we aim to discuss the techniques, tools and challenges for establishing an environment to evaluate an M2M system by simulation before its deployment.

Simulation enables the developer to analyze the performance of an M2M system to observe how the system behaves and ensure that the system meets the performance requirement of the target application. A good simulation framework allows the user to quickly establish a simulation environment which models after the M2M system under design. Through repeated simulation exercises, the developer finds good design points under a set of cost and environmental constraints.

We envision that a good performance evaluation framework provides the following facilities and benefits:

- *Simulation of machines and networks.* It is preferable to execute unmodified target applications in the simulation environment to reduce the extra work needed from the user to set up the simulation. While recent advances of emulation technologies have greatly reduced the difficulties of running unmodified target applications by emulating the instruction set of specific types of processors on PCs, accurate emulation of the I/O and network devices would still require further efforts. For M2M systems, network simulation is also required to study the design of the network and the performance impact from the network. However, traditional network simulation tools used by the network research community, such as NS3 [16], are far too slow to keep up with the virtual machines, as they are designed to model the network in great details. Thus, it remains to be a challenge to provide an integrated machine and network simulation framework for building a large-scale M2M system.
- *Performance/power modeling.* As software is emulated, the execution time and power consumption of each hardware components need to be tracked with performance and power models. Modeling the performance and power consumption of a modern mobile device can be a complex task, which requires the use of simulation tools. The use of performance/power estimation tools allows the user to profile the execution time and the power consumption during the runtime of the application. With proper models, the performance and power consumption on a modern Android smartphone can be estimated fairly accurately [14].
- *Analysis and design automation tools.* While the analysis of an M2M system can be done by tracking the hardware/software events that occur on the machines and the network via simulation, the number of

events can be overwhelming to the user, and it is essential that the tools provide useful analytic information to the users. Similar to those software development tools for single-machine applications, a complete tools kit should enable the user to observe the software behavior, hardware events, and hardware-software interactions as much as possible, for example, profiling an application to reveal the distribution of execution time and the call graph of the application, tracing and correlate specific hardware/software events, and visualizing the message exchanges between machines via a graphical user interface. As we can see later in this chapter, the user can instruct the performance evaluation framework to profile/trace particular hardware/software events in specific functions during and after the simulation. Moreover, it would to assist the user in finding good design choices if the tools can be used in an automated framework for exploring a large design space specified by the user.

We believe that the existing performance evaluation tools are far from sufficient for M2M application development and system design, and the community is in demands of better tools. For that, we took steps to build a framework and will describe the framework to discuss the state of the art for facilitating performance evaluation of M2M systems. The rest of this chapter is structured as follows. Section 9.2 surveys the existing works for developing and evaluating M2M systems. Section 9.3 we discuss the design of such a framework. Section 9.4 we present a case study to illustrate the use of a performance evaluation framework to support the design of a network camera surveillance system. Section 9.5 concludes this chapter.

9.2 Related Works

For WSN systems and applications, various simulation studies and tools have been made available [18, 19, 21, 26]. However, existing studies mostly focused on the network, instead of the machines/devices, since the sensors are composed by low-end processors and simple software with very predictable performance. However, it has become very difficult to properly model and simulate today's large-scale heterogeneous M2M applications because:

- *Complexity of individual machines*: There are powerful devices for executing complicated workloads in an M2M application today. For example, some of the devices that serves as a gateway in an M2M application are running Linux with a high performance embedded processor such as ARM or Intel Atom. The complexity of the machines will only become even more complex in the future.

- *Complexity of machine-to-machine communications*: There may be a variety of network media and the network protocols used in an M2M system. Simulating multiple network media and the network topology is more complicated than simulating a single network medium. The network interface on each individual machine and the software network protocol stack such as TCP/IP also increases the complexity of the machine simulators.

- *Complexity of co-simulation*: Unlike WSN applications, M2M applications usually run across heterogeneous devices, so the simulation environment must be able to support a range of devices that can be used to compose an M2M system. Such a simulation environment can be very complex, and the simulation can be very slow for a large M2M system. In particular, properly synchronizing the execution among heterogeneous simulators without slowing down individual simulators is a challenging issue for the design of co-simulation.

To support the development of M2M systems, simulation frameworks have been proposed by connecting virtual machines (VMs) with a network simulator [2, 5, 8, 24, 29]. Software programs can be executed at hundreds of million instructions per second by a modern virtual machine running on a PC, a speed which is sometimes faster than the actual machine. Thus, the virtual machines are fully capable of executing the entire software stack, including the operating system, to support the development of M2M applications.

While a VM can execute workloads at a variable speed, and the resulted execution time may not faithfully reflect the execution time on the actual machine. Without accurate timing on the execution, the timing of communications between machines may be incorrect. Thus, performance models need to be added for the VMs to model the types of machines in an M2M system and estimate the timing of execution. Based on our VPA framework [14], we can model the low-end and high-end devices in an M2M system by configuring the performance and power models described in the framework. In addition, the VPA framework provides the facilities for profiling and tracing the application running on the VM, which are useful for performance analysis and tuning.

For a simulation environment to simulate a large M2M system, it is important to minimize the overhead of network simulation. A cycle-by-cycle network simulation method would be far too slow in comparison to the speed of virtual machines, so a high-speed event-based simulation method is more desirable. For that, it is possible to parallelize the network simulation based on the topology of the M2M network, so that local communications can be simulated independently. In our Virtual Network Emulation (VNE) framework, we consider the locality of communications and place the VMs which communicate frequently in one host machine. We also introduce methods to reduce the data copy overhead for the network simulator to handle the communication operations between the VMs.

Among the existing works, TOSSIM [18] was capable of simulation for more than one thousand sensor nodes by sacrificing the accuracy of timing estimation. ATEMU [21] supported up to 120 nodes, but it traded its speed for better accuracy. AVRORA [26] aimed to balance the speed and accuracy between TOSSIM and ATEMU and offered tools for analyzing the control flow of the programs and the energy efficiency of the sensor nodes. SUNSHINE [29] was proposed to evaluate hardware designs using cycle-accurate simulators. Thus, it incorporated TOSSIM to simulate the network, SimulAVR to simulate the processor architecture, and GEZEL to simulate the hardware components on the sensor node.

Support for heterogeneity is especially important to M2M applications. For M2M applications, EmStar [8] provides a rich set of software for simulating heterogeneous devices, e.g., 8-, 16- and 32-bit processors, and heterogeneous networks. SEMU [19] used virtual machines to model sensor network and was capable of emulating 1,250 nodes. Qemunet [2] could be used to build a simulation environment by connecting machine emulators and a network simulator. The main focus of Qemunet was to simulate the network traffics generated by M2M applications running on the emulators. Werthmann and Kaschub [24] adopted a similar approach by incorporating virtual machines with an in-house network simulator, *IKR-SimLib*, for analyzing designs of mobile networks.

However, none of the above-mentioned tools contained performance/power models or the tracing facilities to support detailed performance analysis and application development. Thus, we developed a performance evaluation framework with the intention to provide additional assistance to the developers with performance monitoring, event tracing, and visualization.

9.3 Performance Evaluation of M2M Systems

This section describes the framework that we developed to facilitate performance evaluation of an M2M system with two layers of sub-frameworks: one

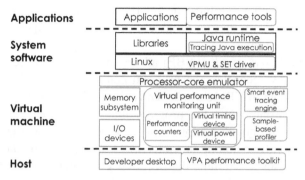

Figure 9.1 Overview of the VPA framework.

for modeling the machines, and one for modeling the network. The machine layer leverages our previous work, *Virtual Performance Analyzer* (VPA) [14], a scheme which builds on the QEMU virtualization software to model the performance and the power consumption of the machines in an M2M system. Section 9.3.1 gives an overview of the VPA framework.

The network layer, called Virtual Network Emulation (VNE), is designed to connect the virtual machines, which will be introduced in Section 9.3.2. Then, the workflow for the simulation of an M2M system is discussed in Section 9.3.3. The mechanism that we used to transmit the data among the emulated nodes is described in Section 9.3.4 and time synchronization that we used in VNE is highlighted in Section 9.3.5. Finally, the way we model the network performance is introduced in Section 9.3.6.

9.3.1 The VPA Framework

Figure 9.1 illustrates the architecture of the VPA framework. The VPA framework has brought the following components to an emulated system. From bottom to top, these components are described below:

- The VPMU is integrated to the popular open-source VM software, QEMU, to support performance monitoring in the VM layer. VPMU includes a set of simulators to model hardware components. The performance counters in VPMU is responsible for collecting hardware and software events. The Virtual Timing Device (VTD) is responsible for estimating the execution time for the emulated system and the Virtual Power Device (VPD) is responsible for estimating power consumption for the hardware components. VPMU exposes a software interface which resembles the interface of the performance monitoring unit (PMU) found on a modern processor to support conventional performance tools.

- A *sample-based profiler* (SBP) allows the user to profile the software execution by recording the program counter along with selected performance data in VPMU *periodically*. The value in the program counter is later translated into the program line or the function name in the system. With enough samples, the profiler may report the distribution of execution time among the software components. The data in VPMU allow the user to examine the occurrence of interested events between samples so as to observe periodic hardware/software behavior in the system.

- The *smart event tracing engine* (SET) allows the user to specify the hardware/software events to be traced and activities to be triggered. The tool covers a wide range of events in the system, from hardware events such as specific machine instructions, cache misses, I/O operations, to software events such as specific Java application functions, libraries, system calls, interrupt handlers, etc. Whenever an event on the trace list is encountered, SET performs the activity specified by the user. For debugging purposes, SET may be used as an *in-circuit emulator* (ICE) for the user to stop the system for specific events and examine the contents in the memory. To analyze a function, SET can record the performance data in VPMU upon entering and leaving the function.

- The *VPMU/SET driver* module is loaded to the Linux kernel for two purposes. First, it exposes VPMU via the *Perfctr* interface to performance tools that are compatible with the interface. Second, it sends symbol tables in the Linux kernel space and the user space to SET, so that SET knows the address for every function in the system and display the function names to the user.

- For *tracing Java applications*, we instrumented the Java runtime library. We managed to make the instrumentation compatible with the just-in-time (JIT) optimizer in the Java VM so that it adds very little overhead to the application execution. The overhead of the tracing mechanism in the Java VM is of very little intrusiveness to performance analysis since we have made sure that the overhead is not accounted by VPMU. It works with SET to trace Java functions by sharing the symbol table of the application running on the Java VM with SET.

- The *performance tools* indicate the support for existing performance tools to take advantage of the VPA framework. For instance, *PAPI* [20]

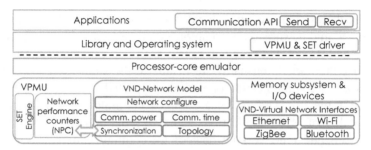

Figure 9.2 Architecture of Virtual Network Emulation framework.

and *PerfSuite* [17] can be ported/modified to work with the VPMU through the Perfctr interface. The *VPA performance toolkit* serves as the frontend for SBP and SET, which aims to minimize the perturbation introduced by the instrumentation and achieve non-intrusive tracing/profiling. The VPA performance toolkit enables the user to profile applications immediately without having to install any patches to the OS kernel.

In our work, we may configure the QEMU and VPA to model various machines in an M2M system. For low-end and mid-range processors, the timing models of VPA are quite fast and accurate. With cache simulation enabled, the simulator can execute tens of ARM million instructions per second, and the performance and power estimates provided by the simulator are approximately within 20% of the actual measurement.

9.3.2 The VNE Framework

Figure 9.2 illustrates the key components of our VNE framework, including a cross-platform *Communication API* and *Virtual Network Device* (VND). In an actual M2M system, the communication libraries and network devices may vary. While it is preferable to execute unmodified target applications in the simulation environment, supporting every available network devices in any machine would be an overwhelming task for us. However, it would put too much the burden on the user to create the virtual network device for the simulation environment. Thus, we provide two methods for the user to set up the M2M communications in the simulation environment.

If the network devices on target machine can be found in the VM, or if the user would like to create detailed models for the network devices, then M2M communications can be performed via the VNDs on the simulated machines. On the other hand, for those M2M systems which are based on a middleware or a common communication library, it may be easier to build a simulation environment with our *Communication API* by abstracting the underlying hardware details in the simulation environment. The abstracted communication operations supported in the current VNE are shown in Table 9.1. The use of a common communication API improves the portability of M2M applications and save programming efforts to set up platform-dependent networking devices for communications. Furthermore, the communication library enables the programmers to test different communication media and choose the best media to transmit the data. There are an increasing number of middleware [25] designed for M2M applications, and it should be quite straightforward to support the middleware with the communication operations in our API.

The implementation of our communication library is linked to the networking devices available on the target system. For example, our communication library is linked to the the *ZigBee* device of a wireless sensor node or the *Ethernet* device in a network router. Each VND has a *Virtual Network Interfaces (VNI)* for protocol dependent configurations of different network interfaces, e.g., environmental parameters deciding the probability of data collision for ZigBee devices. VND also includes a *VND Network Model (VNM)* for keeping general network information, such as the topology of simulated network, and the performance and power models for the VNIs.

The user simply needs to specify the type of network devices for each machine in the simulation environment, and our communication library will model the network devices and the network. This method also simplifies the complexity of simulation and significantly improves the overall simulation speed. Instead of walking through the software network stack and the hardware operations in the device drivers, the communication operations will be bypassed to the specific network interface in the VND to model the performance and the power consumption of the network devices (e.g. ZigBee, Wi-Fi, or Ethernet) as well as the network. The details of the workflow is further discussed in Section 9.3.3

Figure 9.3 illustrates a send-receive process with the VNE framework. Whenever a message is sent from one VM to another via the VNI, the latency of the message is calculated by the *VNM*. The latency is deposited into the *Network Performance Counter* (NPC) of the VPMU, which updates the local clock periodically. Finally, the *Communication Daemon* (CD) in the VND runs as a separate thread for transmitting the data for emulated systems. Since the overhead of message exchanges among the simulated machines is critical

Table 9.1 Application programming interface for M2M communications.

Function prototype	Description
NetworkInit (DeviceID, NetworkStruct, TimingModel)	Register device, initial network buffer and performance counter.
NetworkExit (NetworkStruct)	Release network buffer, reset performance counter.
NetworkSend (ReceiverID, Message, MessageSize, NetworkStruct)	General interface for user send Message with MessageSize to ReceiverID. The NetworkStruct used to indicate network configure (e.g. SOCKET struct for ethernet).
NetworkRecv (SenderID, Message, MessageSize, NetworkStruct)	Receiver data from network buffer.
NetworkWait (NetworkStruct)	Waiting for non-blocking send complete.

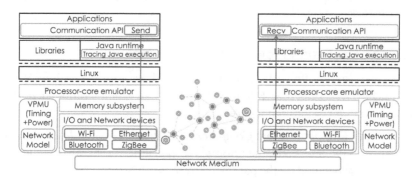

Figure 9.3 Sending and receiving of packets via the VNE framework.

to the efficiency and scalability of the simulation environment, we incorporated the *MSG* library [13], a light-weight, scalable message-passing library which we developed for embedded multcore systems, to the framework to minimize the overhead of M2M communications. The details of the inter-VM communications is further discussed in Section 9.3.4

When each VM executes its local work, it simply executes its programs, accesses its local resources, and advances its local clock. When the VM communicates via the VNE framework, the VND synchronizes the local clocks between the VMs. We implemented a synchronization mechanism based on the distributed synchronization protocol [28] to maintain the time causality of the simulated network events. The details of the synchronization mechanism is further discussed in Section 9.3.5.

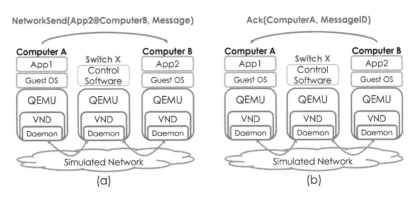

Figure 9.4 An example M2M system emulated by the VNE framework. (a) NodeA send message to NodeB and (b) NodeB send ack back to NodeA.

9.3.3 The Execution Flow of the VNE Framework

The basic execution flow of the VNE framework is described below. We use the M2M system shown as Figure 9.4 to illustrate when Computer *A* sends data to Computer *B* through a network switch.

1. On both *Computer A* and *Computer B*, when the *NetworkInit* function is called in the programs, the library is responsible for initializing the VNI, which is specified by the parameter, *NetworkInterface*. At the same time, the *Communication Daemon*(CD) thread will be created for moving data across the simulated network. As for *Switch X*, the *CD* is created as soon as the system is up.

2. When the sender (*Computer A*) calls the *NetworkSend* function, the parameters will be passed to the VND. The parameters specified in *NetworkStruct* determine whether this function call will be returned immediately (non-blocking) or wait for the result before it can return the data to the application (blocking). In Figure 9.4, a blocking operation is performed, so the function call blocks and waits for the completion of the operation.

3. In the VND, the data is wrapped to form packets and the packet head is shown in Figure 9.5. Note that the *Send Time* is the timestamp when the packet to be sent out from the current node and the *Transmission Time*, which is calculated by the *VNM*, is the estimated delay for sending the data from the current node to the next hop node. The packets in the VND are kept in a queue and will be sent by *CD* to the next hop node, which

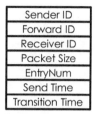

Figure 9.5 Data fields of packet header.

is determined by the routing mechanism of the modelled network, and in this case, the packet will be sent to *Switch X*.

4. When the packet arrives at *Switch X*, it is stored in the *Incoming Header Queue (IHQ)*, which is a FIFO queue to store the incoming packets and maintain their orders. The VND of each node may contain more than several IHQs, one for each connected node. For example, in *Switch X*, there will be two IHQs, one for *Computer A* and the other for *Computer B*. The usage of the *Incoming Header Queue* will be discussed in Section 9.3.4.

   After time synchronization is done between the linked nodes, the *CD* on *Switch X* will fetch *the earliest packet* among the *Incoming Header Queues* and advance the NPC of current node, where the earliest packet is the packet with minimal value of the (*Send Time + Transmission Time*). Based on the updated timing information, the *CD* determines whether the packet should be forwarded or not. In this example, the daemon copies the packet to the IHQ at the *Computer B*, as the device identifier of *Switch X* does not match the destination identifier stored in the packet. In addition, before the data is put to the IHQ, the *Send Time* and *Transmission Time* are updated with the current local clock and the estimated delay for transmit the data from the *Switch X* to the *NodeB*, respectively.

5. On *Computer B*, the *CD* checks if any packet has arrived. If yes, it resolves the payload from the packet, stores the data in the *Network Incoming Data Buffer* and updates the contents (network time consumed in the system) in the NPC. Note that the packets will be examined first before the data can be copied to the *Network Incoming Data Buffer*, which is stored the all the incoming data. The related information is described below. When the *NetworkRecv* function is invoked and blocked waiting for the data, the function will be returned after the library copies the data from the IHQ to the user-space buffer, where the location is specified in

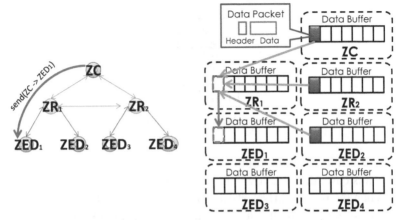

Figure 9.6 Multiple nodes transmit message to the same node.

the function arguments. Finally, an acknowledgement message is sent to the sender node directly by the *CD* with the calculated arrival time, by summing up *Send Time* and *Transmission Time* in the packet.

6. Since the acknowledge message contains the arrival time of the packet at receiver node, the sender can then get the end-to-end communication latency by subtracting the sending time of the packet from its arrival time. In this example, as the blocking operation is considered, the program control will not be returned until the acknowledge message is received and the estimated communication latency is updated to the NPC. It is important to note that the acknowledgement message used for synchronization purpose in the VNE and has nothing to do with the acknowledgement message involved in the modeled network.

Note that for non-blocking operations, the program control is returned as soon as the arguments of the *NetworkSend* function is sent to the VND (in Step 2 above). The programmers can use the *NetworkWait* to check if the function has been finished. Also note that to keep the correct order of the events appeared in the simulated environment, we have implemented an order-controlling mechanism to make sure the events are consumed in the proper order. If the incoming packets are ahead of the local clock at the current node, the packets are not seen by the communication library until the local clock catches up. For instance, in Step 5 above, the *CD* of *NodeB* will be responsible for coping the packet to the *Network Incoming Data Buffer* to indicate the packet is available.

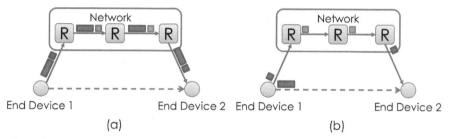

Figure 9.7 (a) General netowork data transmission. (b) Network data transmission on VMs.

9.3.4 Data Transmission among Emulated Nodes

The simulation environment needs to guarantee the functional correctness for the data transmission among the emulated nodes by properly handling the potential race conditions. For example, in Figure 9.6, the ZigBee Coordinator (ZC for short) wants to transmit a message to ZigBee EndDevice (ZED for short) 1. Based on the ZigBee network routing algorithm, message will be passed to ZigBee Router (ZR for short) 1 first, while ZR2 and ZED2 also want to pass messages to ZR1. Thus, a race condition occurs since all three nodes want to send the data to the same node, and we need an efficient mechanism to resolve the race condition. We have surveyed some previous works which attempted to accelerate data transmission between virtual machines [4, 6, 11, 23, 27], where *shared memory* has been utilized as the main communication medium for VM-to-VM communications within a host machine.

In addition to the use of shared memory to speed up data transmission between VMs, the VNE framework also reduces the number of data copies for transmitting packets. Figure 9.7(a) describes the regular packet transmission over the network, where the header and the payload of a packet pass through many intermediate routers. During the simulation, the intermediate routers only need to pay attention to the header information, so the VNE framework only transmits the header through the simulated network, as shown in Figure 9.7(b). The header of the packet goes through the network for updating the time stamp, and the router uses the header information to calculate transmission time and decide the next-hop device. When the header eventually arrives at the destination node, the VNE framework copies the payload data from source node to the destination node. As a result, the payload only needs to be copied once instead of multiple times in the simulated network.

The data structures for the emulated packet transmission are designed for the efficiency and scalabilty of the simulation with a lock-free shared

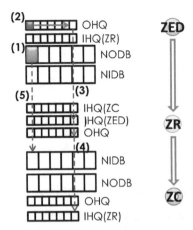

Figure 9.8 An example of data transmission with VND data structure.

memory buffering scheme. We leverage our previous work on the MSG message-passing library [13] and define the following four data structures:

- *Incoming Header Queue (IHQ)* is a FIFO ring buffer used to place the headers of the incoming packets. At the receiver side, there is a HQ for each connected (sender) node. In each HQ, the packets sent from the specific node are sorted by the timing order, which means the top of the queue is the earliest packet in the queue.
- *Outgoing Header Queue (OHQ)* is a FIFO ring buffer for each node used to place the headers of the outgoing packets. As the IHQ, the packets in the OHQ are also ordered by the local clock of the node.
- *Network Incoming Data Buffer (NIDB)* is used to place payloads of the incoming packets which is ready to be read by the *NetworkRecv* function.
- *Network Outgoing Data Buffer (NODB)* is used to keep the outgoing data which is sent by the *NetworkSend* function.

We use Figure 9.8 as an example to illustrate the data transmission with buffers implemented with the aforementioned data structures. In this example, the *End Device (ZED)* wants to transmit a data to the *coordinator (ZC)* through a *router (ZR)*, and the communication flow is the following:

1. *ZED* puts the payload into the NODB and records the entry number on the header.
2. *ZED* puts the header into the OHQ. Since the OHQ is a FIFO ring buffer, the header will be placed in the top of queue.

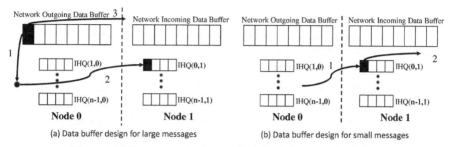

Figure 9.9 The communication method for different payload size.

3. When *CD* on *ZED* detects header in its OHQ, it delivers the header to the IHQ of next device, i.e. *ZR*. In this case, the *ZR* has two IHQs: one for receive data from *ZED*, and one for receive data from *ZC*. In general, the number of IHQs is decided by the number of nodes which are connected to the router, so as to avoid the need for each *CD* thread to acquire a synchronization lock for depositing a header into the IHQ.

4. The *CD* on *ZR* checks all of its IHQs, finds the earliest header, and transmit the header to next device. During the process, the *CD* moves the headers form its IHQs to its OHQ one-by-one according to the arrival time of the headers. Only one OHQ for each device is needed since it is only used by one *CD*. The *CD* on *ZR* then routes the headers to their destination. In this example, the header is copied into the IHQ on *ZC*.

5. When the *CD* on *ZC* detects the arrival of a header in its IHQ, since *ZC* is the final destination of the packet, *ZC* needs to copy the payload from the NODB on the source node *ZED* to its NIDB, and then notifies the application or system process to receive the packet.

The design provides the some advantages for a large-scale M2M system:

- *Lock-free design*: The buffers need not to be protected by synchronization locks when multiple VMs transmit data to the same VM, as each VM has one IHQ to receive the headers from each VM that is connected to it. This saves the expensive lock operations and improves the scalability of simulation performance.
- *Efficient memory utilization*: As mentioned previously, only the header travels through the simulated network, so not only the data copy cost is saved, that the memory requirement is significantly lower since the IHQ/OHQ on the router only need to store the headers. Thus, even with the multiple IHQs in our lock-free design, the memory utilization is still under control for a large-scale M2M network.

- *Optimized for short messages*: For certain WSN applications, the messages are short with very small amount of payload data. Thus, when a message is shorter than a pre-set threshold, the *CD* simply puts the payload in the IHQ/OHQ. As shown in Figure 9.9, instead of putting payload into NOHQ as in (a), the CD puts the both the header and the payload of a packet into IHQ directly as in (b). As a large threshold would adversely increases the data copy cost and the memory requirement, the threshold value is a parameter that depends on the M2M application and system, which can be tuned in practice.

9.3.5 Time Synchronization

During the emulation of an M2M system, the local clocks of the VM nodes need to be synchronized to ensure that the events in the system occur in the correct order. We have surveyed some of the related works [1, 7, 9, 10, 15, 22, 30]. Basically, the existing methods could be divided into categories: *aggressive* and *conservative*. An aggressive (optimistic) synchronization method assumes that every event is safe to proceed on one process without having to consult with the other processes. In case the assumption is wrong, the simulation process is rolled back and re-done with a correct assumption. The rollback may incur the overhead of sending messages to cancel the simulation work done earlier. On the other hand, a conservative synchronization method prevents out of order execution of events. An event is not executed unless it can be guaranteed that no other event with a smaller timestamp will be received.

To reduce the overhead of synchronization, we adopt the distributed synchronization mechanism [28], which synchronize the emulated nodes partially, as illustrated in Figure 9.10. At the gateway (the central node), there are three IHQs, each of them storing the outgoing packets sent by the specific device, and one OHQ to store the packet which is ready to be sent out the gateway. As the top of a IHQ representing the earliest packet in the queue, the *Communication Daemon* finds the earliest packet among the IHQs by comparing the times of the top packets of the IHQs, and copies the earliest packet to the OHQ. Then it checks the local clocks of the surrounding nodes, which are connected to the gateway. If all the local clocks are equal to or faster than the time indicated by the packet, the daemon copies the packet to the IHQ of the receiver node (or the next hop) and updates the NPC of the gateway to honor the fact that the packet has been transferred. In this way, the local clocks are then synchronized since the slower ones can catch up.

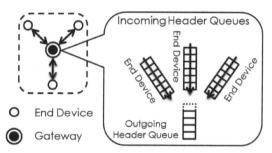

Figure 9.10 Example of time synchronization in the communication system.

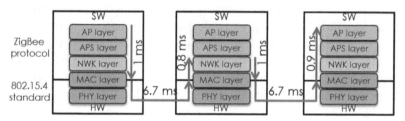

Figure 9.11 Network stack for a Zigbee-based M2M system. The left and right nodes are the end devices, connected to the central node, the router.

On the other hand, to avoid devices running too fast, when a device intends to receive the data with the *NetworkRecv* function, the device has to synchronize its local clock with its neighbor devices.

9.3.6 Modeling the Network Performance

To speed up the simulation speed, the VNE uses the analytical model to estimate the latency for data transmission. A third party network simulators, such as NS2/NS3 [16], can be attached via the VNE framework if needed. As described in Section 9.2, this approach has been adopted by the Qemunet [2], and Werthmann and Kaschub [24].

In our basic performance model, we separate the communication latency into two layers: software latency and hardware latency, which gives the user more flexibility in modeling the dynamics in the network system. For example, the ZigBee network can be modeled as illustrated in Figure 9.11, where the software implementation part and hardware components part refer to the ZigBee protocol stack and IEEE 802.15.4 radio standard, respectively. The idea of the two-layered approach is to obtain the time spent on each part separately by either performance benchmarking or analytical models.

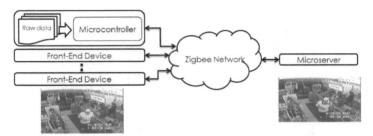

Figure 9.12　A surveillance system with cameras connected to a server via a ZigBee network.

During the emulation, the performance data will be added to the performance counters in VPMU accordingly to reflect the performance impacts of the network operations. The end-to-end communication latency, from the left node to the right node through the router at the central node, can be obtained by aggregating the time spent on the arrows. We have implemented the ZigBee network and used the performance data presented in [3].

9.4　Experimental Results

In this section, due to the length limitation of this chapter, we only present a case study and discuss the experimental results in this case. Our previous results of the VPA framework and the VNE framework can be found in [14] and [12], respectively.

9.4.1　Case Study: Designing Intelligence Surveillance Systems for Intrusion Detection

In this case study, we simulate the surveillance system which consists of a microserver and several front-end devices. Each front-end device contains a camera, a motion sensor, a microcontroller (ARM7TDMI @ 55MHz) and a ZigBee network device, as depicted in Figure 9.12. The front-end device captures an image when the motion sensor detects optical changes. The image is then sent back to the microserver for further processing. As a designer, we would be interested in reducing the power consumption and cost of the system by selecting low-power/low-cost hardware components. At the same time, to ensure that the system can detect intruders in real time, we need to quantify the *end-to-end delay* between the time when the front-end device captured the image and the time when the human faces in the image are identified on the microserver.

Table 9.2 List of experimented system configurations.

Config.	Image resolution & size	Processor type on the microserver
I	640x480/113.5KB	Camera node does all the jobs.
II	640x480/113.5KB	ARM926 @ 260MHz
III	640x480/113.5KB	ARM11 @ 580MHz
IV	640x480/113.5KB	ARM Cortex-A9 @ 1GHz
V	320x240/29.8KB	ARM926 @ 260MHz
VI	320x240/29.8KB	ARM11 @ 580MHz
VII	320x240/29.8KB	ARM Cortex-A9 @ 1GHz

Table 9.3 End-to-end delay (seconds) of the surveillance system with different configurations.

Config.	I	II	III	IV	V	VI	VII
Face detection at camera node	134.8	–	–	–	–	–	–
Image transmission time	7.8	7.8	7.8	7.8	2.1	2.1	2.1
Face detection at microserver	–	32.3	16.4	7.9	8	4.9	1.9
Total time	142.6	40.1	24.2	15.7	10.1	7	4

First, we created an environment to simulate a single camera node and a microserver. We experimented several configurations as listed in Table 9.2. We used the SET tool in the VPA framework to profile each configuration, and the results are shown in Figure 9.13 and Table 9.3. In Config. I, the face detection and data transmission were done by a low-end microcontroller, which took about 142 seconds to locate the potential intruders and sent the JPEG image to the microserver. The delay was simply unacceptable. For the rest of the configurations (II ~ VII), the camera node transmitted the image to the microserver for face detection. The delay depended on two factors: the resolution of the image and the type of processor in the microserver. For example, it took about 7.8 seconds for data transmission and 7.9 seconds for data computation on the microserver in Config. IV. Figure 9.14 shows the call graph of the program running on the microserver to identify the hot functions. In this case, two functions consumed most of the execution time on the microserver: *facedetect()* and *Network_Recv()*.

From our experiences, the results of the face detection application would depend on the image resolutions as the higher resolution provides more image detail. However, the higher image resolution will prolong the data transmis-

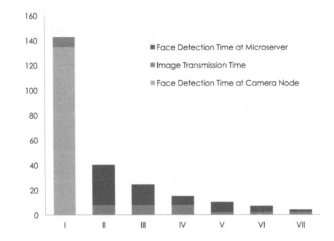

Figure 9.13 Profile data reported by VPA.

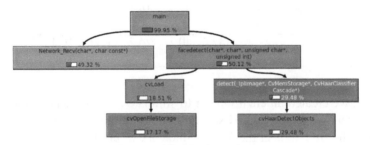

Figure 9.14 Call graph of the microserver.

sion time and processing time. our tool is necessary to examine the possible configurations and obtain both the performance and the results return by the application, so that the system developers can determine whether the overall performance of the novel designs can meet the system specification or not.

As demonstrated here, the VNE/VPA framework enables us to explore the design space of a given M2M application. We can use the framework to study complicated scenarios. For example, what if the number of front-end devices increases in this case study? To certain degree, would the interference and contention on the Zigbee network cause serious packet loss? Would the microserver handle its workload in time? The proposed framework helps answer these questions.

9.5 Conclusion and Future Research Directions

In this chapter, we presented a performance evaluation framework for M2M systems. We show that the framework may be used for modeling a range of machines with adequate accuracies to provide useful performance data to the developers of M2M systems. The use of fast virtual machines and fast virtual network devices is vital to provide a sufficient simulation speed for executing M2M applications. The use of visualization tools is also needed to assist the developer in analyzing the performance data.

In the future, we envision that parallel simulation will be very important when M2M applications and systems become more sophisticated. As modern personal computers and servers are equipped with multicore processors, the simulation environment may be parallelized (multithreaded) to take advantage of the multicore host systems for a order of magnitude of speed-up. As each virtual machine can be executed by an individual processor, the communications and synchronizations between virtual machines can slow down the simulation of the entire M2M system, which is somewhat acceptable within a multicore host, but can seriously hurt the performance if the virtual machines is distributed over a cluster of servers in the cloud. It will be of interest to research on the strategies and schemes for minimizing the impact of inter-VM communications and synchronizations for simulating a large-scale M2M system on the cloud servers.

While the results from our framework seem encouraging, it is obvious that a more complete development environment with performance tools kit is required to accelerate the design of M2M systems in practice. We believe that this is an area which needs continuous research and implementation efforts from the community. As many of the tools mentioned in this chapter, including ours, are open-source, we would like to see more future works devoting to developing new techniques for modeling, simulation, and analysis based on these existing open-source works.

References

[1] K. R. Bisset. An adaptive synchronization protocol for parallel discrete event simulation. In *Proceedings of the 31st Annual Simulation Symposium*, pages 26–33, 1998.

[2] P. Boyko and A. Mazo. Qemunet: An approach to an automated virtualized testbed. In *Proceedings of the 4th International ICST Conference on Simulation Tools and Techniques*, pages 431–438, 2011.

[3] D. Brunelli, M. Maggiorotti, L. Benini, and F. L. Bellifemine. Analysis of audio streaming capability of ZigBee networks. In *Proceedings of the 5th European conference on Wireless sensor networks*, pages 189–204, 2008.

[4] A. Burtsev, K. Srinivasan, P. Radhakrishnan, L. N. Bairavasundaram, K. Voruganti, and G. R. Goodson. Fido: Fast inter-virtual-machine communication for enterprise appliances. In *Proceedings of the 2009 USENIX Annual Technical Conference*, pages 25–25, 2009.

[5] G. Coulson, B. Porter, I. Chatzigiannakis, C. Koninis, S. Fischer, D. Pfisterer, D. Bimschas, T. Braun, P. Hurni, M. Anwander, G. Wagenknecht, S. P. Fekete, A. Kröller, and T. Baumgartner. Flexible experimentation in wireless sensor networks. *Commun. ACM*, 55(1):82–90, 2012.

[6] F. Diakhaté, M. Perache, R. Namyst, and H. Jourdren. Efficient shared memory message passing for inter-vm communications. In *Euro-Par 2008 Workshops – Parallel Processing*, Lecture Notes in Computer Science, volume 5415, pages 53–62. Springer, 2009.

[7] P. Fey, H. W. Carter, and P. A. Wilsey. Parallel synchronization of continuous time discrete event simulators. In *Proceedings of the 1997 International Conference on Parallel Processing*, pages 227–231, 1997.

[8] L. Girod, N. Ramanathan, J. Elson, T. Stathopoulos, M. Lukac, and D. Estrin. Emstar: A software environment for developing and deploying heterogeneous sensor-actuator networks. *ACM Trans. Sen. Netw.*, 3(3), 2007.

[9] A. Grau, K. Herrmann, and K. Rothermel. Efficient and scalable network emulation using adaptive virtual time. In *Proceedings of 18th International Conference on Computer Communications and Networks*, pages 1–6, 2009.

[10] Z. He, X. Yao, and X. Zhao. Discrete-event model of time synchronization. In *International Symposium on Computational Intelligence and Design*, volume 1, pages 340–343, 2011.

[11] W. Huang, M. J. Koop, Q. Gao, and D. K. Panda. Virtual machine aware communication libraries for high performance computing. In *Proceedings of the 2007 ACM/IEEE conference on Supercomputing*, pages 9:1–9:12, 2007.

[12] Shih-Hao Hung, Chun-Han Chen, and Chia-Heng Tu. Performance evaluation of machine-to-machine (M2M) systems with virtual machines. In *Proceedings of 15th International Symposium on Wireless Personal Multimedia Communications*, pages 159–163, 2012.

[13] Shih-Hao Hung, Po-Hsun Chiu, and Chi-Sheng Shih. Building a scalable and portable message-passing library for embedded multicore systems. In *Proceedings of the 2011 ACM Symposium on Research in Applied Computation*, pages 31–37, 2011.

[14] Shih-Hao Hung, Tei-Wei Kuo, Chi-Sheng Shih, and Chia-Heng Tu. System-wide profiling and optimization with virtual machines. In *Proceedings of the 17th Asia and South Pacific Design Automation Conference*, pages 395–400, 2012.

[15] D. Jin, Y. Zheng, H. Zhu, D. Nicol, and L. Winterrowd. Virtual time integration of emulation and parallel simulation. In *Proceedings of the 2012 ACM/IEEE/SCS 26th Workshop on Principles of Advanced and Distributed Simulation*, pages 201–210, 2012.

[16] K. Fall and K. Vardhan. The Network Simulator – NS-2. http://www.isi.edu/nsnam/ns.

[17] R. Kufrin. Perfsuite: An accessible, open source performance analysis environment for Linux. In *Proceedings of the 6th International Conference on Linux Clusters: The HPC Revolution 2005*, 2005.

[18] P. Levis, N. Lee, M. Welsh, and D. Culler. TOSSIM: Accurate and scalable simulation of entire TinyOS applications. In *Proceedings of the 1st International Conference on Embedded Networked Sensor Systems*, pages 126–137, 2003.

[19] Shih-Hsiang Lo, Jiun-Hung Ding, Sheng-Je Hung, Jin-Wei Tang, Wei-Lun Tsai, and Yeh-Ching Chung. SEMU: A framework of simulation environment for wireless sensor networks with co-simulation model. In *Proceedings of International Conference on Grid and Pervasive Computing*, pages 672–677, 2007.

[20] S. Moore, D. Terpstra, K. London, P. Mucci, P. Teller, L. Salayandia, A. Bayona, and M. Nieto. PAPI deployment, evaluation, and extensions. In *DOD_UGC'03: Proceedings of the 2003 DoD User Group Conference*, pages 349–353, 2003.

[21] J. Polley, D. Blazakis, J. McGee, D. Rusk, and J. S. Baras. ATEMU: A fine-grained sensor network simulator. In *Proceedings 2004 First Annual IEEE Communications Society Conference on Sensor and Ad Hoc Communications and Networks*, pages 145 – 152, 2004.

[22] F. Ring, G. Gaderer, A. Nagy, and P. Loschmidt. Distributed clock synchronization in discrete event simulators for wireless factory automation. In *Proceedings International IEEE Symposium on Precision Clock Synchronization for Measurement Control and Communication*, pages 103–108, 2010.

[23] J. Sonnek, J. Greensky, R. Reutiman, and A. Chandra. Starling: Minimizing communication overhead in virtualized computing platforms using decentralized affinity-aware migration. In *Proceedings of the 39th International Conference on Parallel Processing*, pages 228–237, 2010.

[24] T. Werthmann, M. Kaschub, C. Blankenhorn, and C. M. Mueller. Td(11)01038 approaches for evaluating the application performance of future mobile networks. COST IC1004, 2011.

[25] T. Takahashi, S. Sumimoto, A. Hori, H. Harada, and Y. Ishikawa. PM2: High performance communication middleware for heterogeneous network environments. In *Proceedings ACM/IEEE 2000 Conference Supercomputing*, page 16, 2000.

[26] B. L. Titzer, D. K. Lee, and J. Palsberg. Avrora: Scalable sensor network simulation with precise timing. In *Proceedings Fourth International Symposium on Information Processing in Sensor Networks*, pages 477–482, 2005.

[27] Jian Wang, Kwame-Lante Wright, and Kartik Gopalan. Xenloop: A transparent high performance inter-vm network loopback. In *Proceedings of the 17th International Symposium on High Performance Distributed Computing*, pages 109–118, 2008.

[28] Y. Wen, R. Wolski, and G. Moore. Disens: Scalable distributed sensor network simulation. In *Proceedings of the 12th ACM SIGPLAN symposium on Principles and Practice of Parallel Programming*, pages 24–34, 2007.

[29] J. Zhang, Y. Tang, S. Hirve, S. Iyer, P. Schaumont, and Y. Yang. A software-hardware emulator for sensor networks. In *Proceedings 2011 8th Annual IEEE Communications Society Conference on Sensor, Mesh and Ad Hoc Communications and Networks*, pages 440–448, 2011.

[30] Y. Zheng and D. M. Nicol. A virtual time system for openvz-based network emulations. In *2011 IEEE Workshop on Principles of Advanced and Distributed Simulation*, pages 1–10, 2011.

Author Index

Subject Index

About the Editors

Fabrice Theoleyre received his M.Sc and Ph.D. degrees in electrical and computer engineering both from INSA of Lyon (France) in 2003 and 2006 respectively. He joined the CNRS as a research scientist in 2007. After having spent two years in the Grenoble Informatics Laboratory, he joined the university of Strasbourg (ICUBE) in 2009. He served in the TPC of various international conferences (e.g. PIMRC, VTC, IWCMC, WIMOB, WPMC) and is associate editor for *IEEE Communications Letters* and the *Journal of Networks*. His research interests comprise sensor, ad hoc and mesh networks, including distributed algorithms, experiments, MAC layer and mobility management.

Ai-Chun Pang received the B.S., M.S. and Ph.D. degrees in Computer Science and Information Engineering from National Chiao Tung University, Taiwan, in 1996, 1998 and 2002, respectively. She joined the Department of Computer Science and Information Engineering (CSIE), National Taiwan University (NTU), Taipei, Taiwan, in 2002. Currently, she is a Professor in CSIE and Graduate Institute of Networking and Multimedia of NTU, Taipei, Taiwan. Her research interests include wireless networking, mobile computing, and performance modeling.